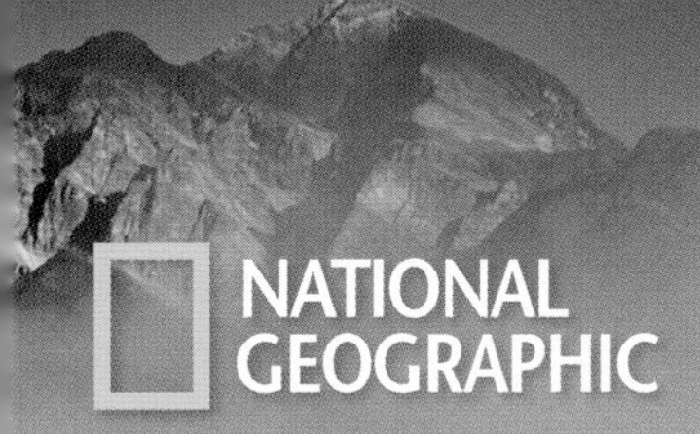

POCKET GUIDE TO THE Rocks & Minerals OF NORTH AMERICA

SARAH GARLICK

NATIONAL GEOGRAPHIC
WASHINGTON, D.C.

CONTENTS

Introduction 6

1 **Minerals** 12

2 **Rocks** 56

3 **Fossils** 100

4 **Structures & Landforms** 120

Natural Regions of North America 172

Further Resources 175

About the Author • Acknowledgments 176

Illustrations Credits 177 | Index 179

Rocks & Minerals

Earth's Beauty and Power

With all the bustle and show of living things, it can be easy to overlook the ground beneath our feet. But Earth's rocks and minerals are among the most fascinating aspects of the natural world. They are the record of billions of years of Earth's activity. Within their structures, from the smallest crystal to the largest mountain face, the beauty and dynamism of our planet are revealed.

What Are Minerals?

Minerals are the solid substances that make up rocks. There are thousands of different minerals on Earth, some rare and valuable like diamond and gold, others, such as quartz, as common as beach sand. Minerals are defined by a few characteristics: They are solids. They are naturally occurring. They have an ordered atomic arrangement. They have a well-defined chemical composition.

Most minerals exist as small, irregular grains within rocks. Occasionally, isolated minerals have beautiful geometric shapes and smooth planar surfaces. These shapes and surfaces are crystals, and they are the outward expressions of the mineral's internal atomic arrangement. Crystals form when a mineral has sufficient time and space in which to grow.

Gems, or gemstones, are materials cut and polished, used for jewelry or other decorative items. Most gems, unless they are synthetically produced, are minerals with gem-specific names. For example, amethyst is a purple variety of quartz, and emerald is a green variety of beryl.

Identifying minerals can be both fun and challenging. It takes close observation, a few basic tests, and the process of elimination—and it definitely gets easier as you gain experience. You don't need many tools: A hand lens and a small hammer will get you started. Mineral identification kits containing streak plates, hardness tools, and a small bottle of diluted acid are helpful.

In a microscopic view of granite, polarized light interacts with mineral grains, revealing brilliant colors.

Three Types of Rocks

Rocks are the solid materials that make up the planet. There are three basic types: igneous, sedimentary, and metamorphic.

Igneous rocks form when molten material, called magma, cools and solidifies. Igneous rocks can be either intrusive, meaning the magma solidified underground, or extrusive, meaning the magma erupted onto the Earth's surface before solidification. (Erupting magma is called lava.) Intrusive igneous rocks cool slowly, allowing minerals to grow into visible grains that form an interlocking, crystalline texture. Extrusive rocks cool quickly, resulting in a fine-grained volcanic texture. Some igneous rocks have two grain sizes with large, well-formed crystals surrounded by a groundmass of finer grains. This is called a porphyritic texture, and the large grains are known as phenocrysts.

In addition to texture, igneous rocks are categorized by chemistry. Mafic rocks are rich in iron and magnesium (minerals such as biotite and hornblende) and poor in silica (such as quartz and feldspar), and they tend to be dark in color. Felsic rocks are light in color, containing more silica-rich minerals and fewer that are rich in iron and magnesium.

Sedimentary rocks form from an accumulation of sediments. The sediments can be preexisting minerals or rock fragments—collectively called clasts—that have been transported by wind, water, or ice; or they can be chemically precipitated or biologically generated in place, as with many types of limestones. During accumulation, sediments become compacted by the weight of material above. Finally, groundwater or other fluids passing through the formation

KEY MINERAL PROPERTIES

- **Hardness:** See inside back cover.
- **Streak:** Streak is the color of the mineral when it is ground into a fine powder. Many mineral identification kits contain small white porcelain plates called streak plates. Scratching an edge of a specimen against this plate will produce a streak. If the mineral is clear, or if it is harder than the porcelain plate, the streak will be white.
- **Crystal habit or aggregation:** The form of a mineral specimen can be helpful for identification. If the specimen has well-developed crystal faces, the shape of the crystal, known as its habit, can be diagnostic. Many specimens, however, do not have well-developed crystal faces and are instead aggregates of small mineral grains. A variety of terms describe crystal habit and state of aggregation:
  - **Prismatic** means that the crystal is long in one direction with well-developed faces. Some prismatic crystals have blunt ends; they can also be topped with pyramid shapes.
  - **Columnar** means that the crystal is long in one direction and resembles the shape of a rounded column.
  - **Tabular** and **platy** habits refer to crystals that have two dimensions relatively equal in length and one that is short.
  - **Bladed** habit refers to crystals that are both elongate and flat, like a blade.
  - **Fibrous** means that the crystal grows in the shape of threads or filaments.

precipitates a cement between the sediment grains—commonly silica or calcium carbonate—hardening the sediment pile into stone. Clastic sedimentary rocks such as conglomerates, sandstones, and mudstones are classified by grain size.

Metamorphic rocks form when preexisting rocks are transformed by heat and pressure, a process called metamorphism. This process involves physical and chemical changes that occur

Massive means that the mineral specimen does not have crystal faces.

Granular describes aggregates of mineral grains that are all approximately the same size and dimension.

Compact specimens are very fine grained so individual grains cannot be distinguished.

Mammillary, from the Latin *mamma,* meaning "breast," describes specimens that are smooth and rounded.

Botryoidal is the term for specimens that are smooth and globular, with smaller spherical shapes than mammillary specimens.

+ **Luster:** The way light is reflected from the surface of a mineral is called luster. Terms describing luster include metallic, glassy or vitreous, pearly, waxy, earthy, and dull. One mineral can display different types of luster depending on its crystal faces or habit.

+ **Color:** Color is an important property used in mineral identification, but it can be misleading because most minerals come in a variety of hues.

+ **Cleavage & fracture:** Some crystals tend to break along one or more smooth, flat surfaces known as cleavage planes. The orientation of these planes is helpful for identifying minerals. Breaks along irregular surfaces instead of cleavage planes are called fractures. Conchoidal fracture is a type of fracture with a smooth, scooped, or curved pattern, common in quartz and obsidian.

Crystal clear water shimmers over stones at Kinta Lake, Glacier National Park, Montana.

in the solid state—changes such as the growth of new minerals, and the physical rotation or recrystallization of existing minerals. The parent rock that becomes a metamorphic rock is called a protolith. Limestone, for example, is the protolith for the metamorphic rock marble, and granite is a protolith for some gneisses. Metamorphism occurs when rocks are subjected to high pressures and/or temperatures—conditions that happen during movements of Earth's plates. When plates collide and push up large mountain ranges, for example, rocks are squeezed and heated in the thickening crust. Metamorphism also occurs in shallow regions of the crust by the heat given off of large magmatic intrusions.

Landforms & Structures

The study of the Earth is not just about identifying rocks and minerals. In fact, what most people notice first are not individual rocks, but rather the shapes and textures of the land—the natural features that make up the landscape. Why does a valley have a certain profile, for example, or why is one coastline rimmed with cliffs and another gentle beaches? These are landforms, and they can tell us a great deal about how our planet works. Most landforms are the result of a combination of features and processes—the structures of the rocks combined with the effects of weathering and erosion. Weathering is the breakdown of rocks by either physical or chemical means. Erosion is the removal and transport of rock materials by the energy of wind, water, or ice.

Fossils

Fossils are the preserved remains or impressions of ancient living things. Fossils form when organisms are buried in sediment and preserved via processes like compaction, chemical alteration, or replacement with minerals. Fossils are extremely important because they are our primary clues about past life-forms and previous environments that existed on our planet—the key to our understanding of evolution. The best places to look for fossils are areas where sedimentary rocks are exposed along road cuts, hillsides, or along streams and rivers. Fossils are most common in fine-grained sedimentary rocks such as mudstones and shales, and marine sedimentary rocks such as limestones.

Use Caution

Collecting rocks, minerals, and fossils is an exciting hobby, but one that carries significant risks and responsibilities. Be sure to check the status of a prospective site beforehand. Gain permission if it is private land, or conform to the rules and regulations regarding prospecting and collecting if the land is public. Information can be obtained from state geological surveys, state parks, the U.S. Forest Service, and the Bureau of Land Management. Always put your safety first. Wear protective gear, especially when using hammers, chisels, or other digging equipment, and be aware of risks like rockfall and unstable slopes. Never enter closed or abandoned mines.

Quartz

Class: Silicate Chemical formula: SiO_2

Quartz is one of the most common minerals in the Earth's crust, found in many sedimentary, igneous, and metamorphic rocks. Collectible varieties include amethyst and smoky quartz.

KEY FACTS

Well-formed crystals are hexagonal prisms with pyramid-shaped tops; quartz comes in various colors, usually clear, gray, or milky white; transparent to translucent.

+ hardness: 7

+ streak: White

+ locations: This is a common rock-forming mineral; gem-quality crystals are found in veins and granitic pegmatites.

Quartz, an important building block of the Earth's crust, consists of silica and oxygen atoms linked in a strong, three-dimensional framework. Quartz is distinguished by its hardness and, in well-formed crystals, its hexagonal-shaped prisms with pyramidal ends. Traces of other elements substituting for silicon create different varieties. Iron in the structure forms yellow quartz crystals known as citrine, or purple crystals known as amethyst. Smoky quartz contains trace amounts of aluminum. Quartz has a vitreous or greasy luster and no cleavage. Microcrystalline forms and aggregates have a conchoidal fracture.

Chalcedony

Class: Silicate Chemical formula: SiO_2

Chalcedony is a variety of quartz made up of crystals so small they cannot be seen with the naked eye. It comes in beautiful colors and patterns, some valued as semiprecious stones.

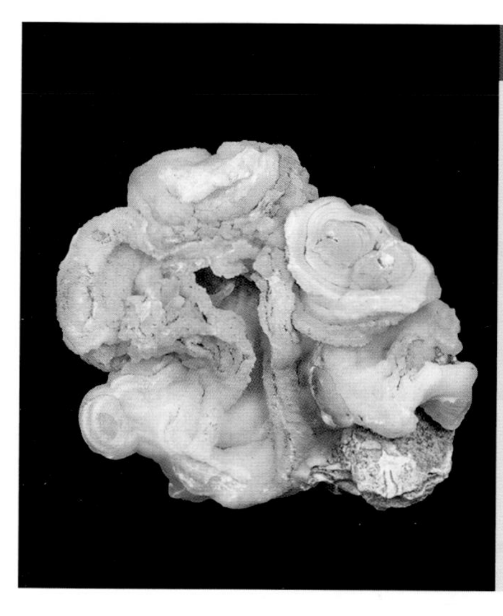

KEY FACTS

This attractive mineral is hard and compact, translucent to opaque, with a waxy or vitreous luster and a conchoidal or uneven fracture.

- **hardness: 6–7**
- **streak: White**
- **locations: Cavities in volcanic rocks and as nodules in sedimentary rocks; clasts of chalcedony can be found in some ocean beach and river gravels.**

Chalcedony is microcrystalline or cryptocrystalline quartz. Microcrystalline crystals are so small that they can be identified only under a microscope; cryptocrystalline crystals, from the Greek *krypto,* meaning hidden, are so small that an optical microscope cannot resolve them. Chalcedony can form in rock cavities, most often in fine-grained volcanic rocks. It is commonly found in globular forms—botryoidal or mammillary shapes—and as the outer shell of geodes and the material replacing plant parts in petrified wood. Collectible varieties include agate, bloodstone, carnelian, onyx, and petrified wood.

Agate

Class: Silicate Chemical formula: SiO_2

A concentrically banded form of chalcedony, agate is commonly found filling rounded cavities in volcanic rocks. Cut and polished agates are sold as ornaments and used in jewelry.

KEY FACTS

This microcrystalline to cryptocrystalline quartz is banded typically parallel to its cavity; wide range of natural colors; many bright colors of commercial specimens are artificial.

+ hardness: 6–7

+ streak: White

+ locations: Rounded nodules in various rock types; cavities in volcanic rocks; eroded agates in river deposits

Agate is known for its beautiful colors and concentric bands, which are often enhanced by dyes and polish in commercial samples. Most agates form from silica-rich fluids that have filled open pockets or seams in volcanic rocks. The silica is deposited in bands of tiny, parallel fibrous crystals around the cavity's inside wall. During crystallization, impurities collect along the bands, creating alternating colors and rings. Look for well-developed crystals (often amethyst or smoky quartz) at centers of some agate nodules. Collectors use the term "agate" for nonbanded chalcedony such as moss agate and flame agate.

Opal

Class: Silicate Chemical formula: $SiO_2 \cdot nH_2O$

Opal is a form of cryptocrystalline quartz with water present. It is made of tiny spheres packed together that interact with light to create a distinctive "opalescent" rainbow shimmer.

KEY FACTS

Variable in color, opal exists as compact masses, crusts, and veins with vitreous or pearly luster.

+ **hardness: 5.5–6**

+ **streak: White**

+ **locations: Near geysers and hot springs; also as pseudomorphs of marine shells or wood, meaning it has replaced the original material while retaining the item's shape and form**

Opal is hardened silica gel in the form of tiny, tightly packed spheres with a significant amount of water (5 to 10 percent) in its pores. It is fragile and can easily crack and lose its water content over time, which diminishes its opalescence. Opal is usually white, but can also be green, brown, black, yellow, or gray due to various impurities such as iron and manganese. Opal is deposited by silica-rich waters at relatively low temperatures. It forms in small cavities in basalt, in cracks and small veins, and as pseudomorphs of wood and shells. The word "opal" comes from Greek *opallios,* meaning "precious stone."

Jasper

Class: Silicate Chemical formula: SiO_2

Jasper is a granular form of microcrystalline quartz, distinguished by its opacity, rich colors, and abundant impurities. It is typically deep red, yellow, or brown due to traces of iron oxide.

KEY FACTS

With hues of rich red, yellow, or brown, jasper is opaque and can have an angular, broken appearance or fine color banding.

- **hardness: 6–7**
- **streak: White**
- **locations: In association with various sedimentary rocks, banded iron formations, and fault zones; jasper pebbles and cobbles also occur in stream deposits.**

Jasper is evenly colored or has banding or brecciation, an amalgamation of angular, broken shards. Polished jasper is traded as an ornamental semiprecious stone. Banded jasper differs from agate by its opacity (agate is typically translucent or semitranslucent). Classification of chalcedony, agate, jasper, chert, and flint is inconsistent. Agate and jasper are commonly considered to be varieties of chalcedony, which is cryptocrystalline or microcrystalline quartz. Chert and flint are often classified as sedimentary rocks composed of microcrystalline quartz. Jasper can also be classified as a colorful form of chert.

Chert/Flint

Class: Silicate Chemical formula: SiO_2

Chert, also known as flint, is made of cryptocrystalline or microcrystalline quartz. Chert occurs as layered formations or as nodules in marine sedimentary rocks.

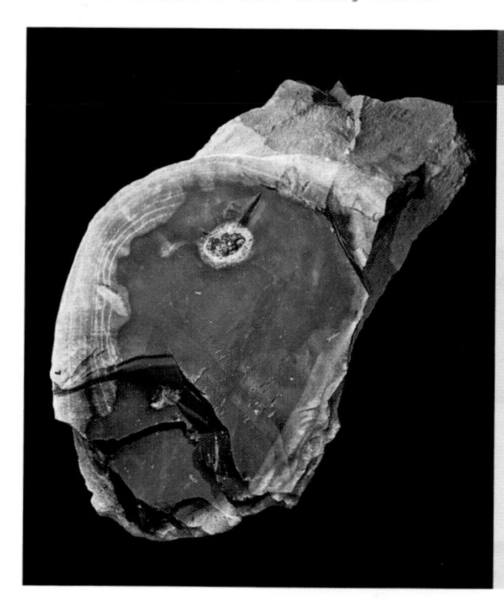

KEY FACTS

Nodules and beds are opaque, hard, and dense masses with conchoidal or irregular fracture; white and gray varieties are often called flint.

- **hardness: 7**
- **streak: White**
- **locations: Found as nodules in chalk and other marine sedimentary rocks; also in association with Precambrian banded iron formations**

Chert, also known as flint, is a hard, compact form of cryptocrystalline or microcrystalline quartz primarily marine in origin. Chert is commonly gray, white, yellow, black, or brown. It forms irregular sedimentary bands or layers, or nodular masses, in fine-grained limestones and chalk. It forms by accumulation of silica, sometimes from organic sources such as needlelike structures in marine sponges. Some cherts contain small fossils. Flint is the name commonly used to describe nodular chert. Chert is usually classified as a rock, but it is included here with other forms of microcrystalline quartz.

Feldspars

Class: Silicate Chemical formula: $XAl_{(1-2)}Si_{(3-2)}O_8$

Feldspars are a group of minerals made of silicon, oxygen, and aluminum, plus various amounts of sodium, calcium, and/or potassium. They are common constituents of the Earth's crust.

KEY FACTS

These common rock-forming minerals are light in color and are often blocky. They display two cleavage planes that intersect at an angle of approximately 90 degrees.

- **hardness: 6–7**
- **streak: Most feldspars have a white streak.**
- **locations: Widespread across all rock types; common in intrusive igneous rocks**

Feldspars are building blocks of many rocks, especially granites and other intrusive igneous rocks. They are also found in clastic sedimentary rocks and as grains in metamorphic rocks. A key component of the Earth's crust, feldspars are also found in meteorites and rocks from the moon. Feldspars come in various colors, commonly white, pink, gray, or other light shades. They are distinguished from other minerals by their hardness and cleavage planes. Feldspar is mined for use in glassmaking and ceramics, and is used in products such as fiberglass, floor tiles, tableware, sinks, toilets, and paints.

Potassium Feldspars

Class: Silicate Chemical formula: $KAlSi_3O_8$

The potassium feldspars are a subgroup of feldspar minerals that contain potassium in their structures. They are important components of igneous rocks.

KEY FACTS

These feldspars with salmon pink grains are commonly found in granite; however, they can be white, gray, or yellow, and are often difficult to distinguish from plagioclase.

+ **hardness:** 6–7
+ **streak:** White
+ **locations:** Granite and other intrusive igneous rocks; well-formed crystals in granitic pegmatites

Potassium feldspars (also known as K-feldspar or K-spar) include microcline, orthoclase, and sanidine. They form short tabular or prismatic crystals. Microcline is milky white, pink, or blue-green and found in metamorphic and igneous rocks. Well-formed microcline crystals occur in granitic pegmatites. Orthoclase is common in granite and other intrusive igneous rocks; sanidine is a high-temperature form of potassium feldspar common in rhyolite and other extrusive igneous rocks. Orthoclase with fine intergrowths of sodium feldspar (albite) is known as moonstone and displays a unique iridescent luster.

Plagioclase

Class: Silicate Chemical formula: $NaAlSi_3O_8$ (albite)–$CaAl_2Si_2O_8$ (anorthite)

Plagioclase is a common mineral and a primary component of many igneous rocks including granite.

KEY FACTS

Commonly white or gray, the crystals have two cleavage planes that intersect at approximately 90 degrees; fine parallel striations on one cleavage face distinguish it from potassium feldspar.

- **hardness: 6–7**
- **streak: White**
- **locations: Metamorphic rocks including gneisses and amphibolites; common in igneous rocks**

Plagioclase is the name for a group of feldspars that have a range of compositions from a pure sodium variety (albite) to a pure calcium variety (anorthite). The plagioclase feldspars include albite, oligoclase, andesine, labradorite, and anorthite. It can be difficult to distinguish between the different types of plagioclase in the field. Plagioclase is usually white or light gray, but can also be dark gray, bluish white, or greenish white. Smooth cleavage planes have a vitreous luster. Like other feldspars, plagioclase is distinguished by its hardness and cleavage planes. Single crystals are rare.

Labradorite

Class: Silicate Chemical formula: $(Ca,Na)(Al,Si)_4O_8$

Labradorite is one of the plagioclase feldspars, containing both calcium and sodium in its structure. Labradorite is known for its large crystal masses with distinct iridescence.

KEY FACTS

Although it is commonly dark blue or gray, some varieties of labradorite can be light in color. The cleavage planes display a diagnostic iridescent color play, called schiller.

+ hardness: 6–7

+ streak: White

+ locations: Found in large crystalline masses in anorthosite; also occurs as a component of gabbro

Labradorite crystals display a striking iridescent color on their cleavage surfaces, known as labradorescence, making it one of the most beautiful of the feldspars. Labradorescence is caused by the interaction of visible light with microscopic layers within the labradorite crystal structure. Labradorite commonly forms large aggregate masses; individual crystals are tabular. Slabs of cut and polished labradorite are used as a decorative facing material of many buildings. Labradorite is named after its type area in northeastern Labrador. It is also common in the anorthosites of the Adirondacks in New York.

Nepheline

Class: Silicate Chemical formula: $(Na,K)AlSiO_4$

Nepheline is the most common feldspathoid, a group of minerals similar to feldspars but undersaturated in silica. It can be found in silica-poor igneous and metamorphic rocks.

KEY FACTS

This silicate is distinguished from feldspars by lack of cleavage, and is distinguished from quartz by hardness (nepheline is softer); quartz is also more resistant to weathering.

- **hardness: 5.5–6**
- **streak: White**
- **locations: Alkalic volcanic rocks and syenites; cannot coexist in a rock with quartz**

Nepheline is a member of the feldspathoids, with chemistries similar to potassium feldspars but with less silica. It is an important constituent of silica-poor igneous rocks such as phonolites and syenites (that is, nepheline syenite). Nepheline is usually white or gray with a greasy luster. It is susceptible to weathering and, thus, can appear pitted or partially dissolved. Well-developed crystals are rare, and often have a prismatic habit with a hexagonal cross section. Nepheline can be difficult to distinguish from quartz but is slightly softer and cannot coexist with quartz in the same rock.

Olivine

Class: Silicate Chemical formula: $(Mg,Fe)_2SiO_4$

Olivine is a green mineral that is part of many iron- and magnesium-rich igneous rocks such as basalt and peridotite. It is the primary component of the Earth's upper mantle.

KEY FACTS

The name refers to its olive green color. It has a vitreous luster and hardness; well-formed crystals are rare.

+ hardness: 6.5–7

+ streak: White or colorless

+ locations: Found in basalts and other mafic volcanic rocks as isolated inclusions; a primary constituent of ultramafic rocks like peridotites and dunites

Olivine grains, with a vitreous luster, translucence, and green or yellow-green color, can look like fragments of green sea glass. Transparent or translucent green crystals are known as the gem peridot. Olivine has variable amounts of iron and magnesium. Rare, pure iron varieties are called fayalite; pure magnesium varieties are called forsterite. Only fayalite coexists with quartz. Olivine is a major constituent of rocks from the upper mantle and oceanic crust. It is found in mafic and ultramafic igneous rocks, and less commonly in metamorphic rocks including marbles. Hawaii's green beaches consist of olivine.

Garnet

Class: Silicate Chemical formula: $X_3Y_2(SiO_4)_3$

Garnets are colorful minerals common in igneous and metamorphic rocks. Individual types of garnet are defined by their chemistry—how different elements fit into their crystal structure.

KEY FACTS

Although usually deep red, garnets can be other colors. The mineral can be distinguished by its hardness and its well-formed 12-sided crystals.

- **hardness: 6.5–7.5**
- **streak: White or colorless**
- **locations: Most often occurs as small crystals in schists; also found in granitic rocks, pegmatites, and marbles**

Garnet types include almandine, pyrope, spessartine, and grossular. Specimens commonly have well-developed crystal faces. In schists and other rocks, the symmetrical, colorful crystals are distinctive, especially when surrounded by flattened grains of other minerals. The most recognizable garnets are deep red with a vitreous luster and high hardness, but they can be orange, yellow, green, blue, purple, pink, brown, and black. Garnets are used as gems and industrial abrasives. Their presence can give geologists a clue to where the rock came from; for example, some garnets indicate origins in the Earth's mantle.

Andalusite

Class: Silicate Chemical formula: Al_2SiO_5

Andalusite is one of three aluminosilicate minerals with the same chemistry but different structures: andalusite, kyanite, and sillimanite. It is found in metamorphic rocks.

KEY FACTS

Prismatic pink, violet, green, or gray crystals have nearly square cross sections; the crystals can be twinned. Andalusite has one perfect cleavage plane and an uneven or conchoidal fracture.

+ hardness: 6.5–7.5

+ streak: White to colorless

+ locations: Can be found in schists and other aluminous metamorphic rocks

Three aluminosilicate minerals have the same chemistry but different crystal structures, known as polymorphs: andalusite, kyanite, and sillimanite. Each polymorph forms under different pressure and temperature conditions, so finding one of these minerals is an important clue about the history of the rock. Andalusite forms under high temperatures at relatively low pressures. Nearly square cross sections of prismatic andalusite crystals are distinctive, but andalusite can also have a massive or compact form. A variety of andalusite called chiastolite contains dark inclusions that create the shape of a cross.

Kyanite

Class: Silicate Chemical formula: Al_2SiO_5

Kyanite is an aluminosilicate mineral commonly forming long tabular or bladed crystals. It is an indicator of medium- to high-pressure metamorphism.

KEY FACTS

Although it is best known for its cyan-blue color, other colors exist in kyanite, including clear, white, gray, and yellow.

+ hardness: 6.5–7 perpendicular to long axis; 4–5 parallel to long axis

+ streak: White

+ locations: Found in mica schists and gneisses; also occurs in some hydrothermal quartz veins

Kyanite is the medium- to high-pressure polymorph of the three aluminosilicates: andalusite, kyanite, and sillimanite. Kyanite specimens commonly display elongate, bladed, or tabular crystals. Kyanite also displays two different levels of hardness corresponding to different regions of the crystal. They are hard (hardness of 6.5 to 7) in the direction perpendicular to the long axis of the crystal and soft (hardness of 4 to 5) parallel to that axis. Kyanite is often found in association with staurolite and garnet in metamorphic rocks. Kyanite is used in a variety of refractory and abrasive products.

Sillimanite

Class: Silicate Chemical formula: Al_2SiO_5

Sillimanite is one of the three aluminosilicates, along with andalusite and kyanite, that have the same chemistry but different structures. Sillimanite grains are often fibrous or needlelike.

KEY FACTS

Characteristic forms of sillimanite include colorless, white, or gray needles as well as fibrous masses. Variations are brownish or blue-green.

+ **hardness:** 7

+ **streak:** White

+ **locations:** Just as its close chemical relatives andalusite and kyanite, this aluminosilicate polymorph occurs in mica schists and gneisses.

Sillimanite is the high-temperature polymorph of the three aluminosilicates: andalusite, kyanite, and sillimanite. It is commonly colorless or white, but can also be brown, yellowish brown, or blue-green. Sillimanite crystals are usually prismatic and fibrous with vitreous luster. Fibrous varieties are also known as fibrolite. It is found in aluminous metamorphic rocks, commonly associated with corundum, cordierite, andalusite, and kyanite. It is the state mineral of Delaware, where it is common in the schists of the Delaware Piedmont, in places as large masses and rounded boulders.

Staurolite

Class: Silicate Chemical formula: $Fe_2Al_9Si_4O_{22}(OH)_2$

Staurolite is a hard, medium- to dark-colored mineral found in some metamorphic rocks. Specimens often have well-developed crystal faces creating prismatic shapes.

KEY FACTS

Two crystals of staurolite are often found grown together, some at right angles. Right-angled "fairy crosses" or "fairy stones" are thought by some people to bring good luck.

- **hardness: 7–7.5**
- **streak: White to gray**
- **locations: Like andalusite and kyanite, it can be found in mica schists and gneisses.**

Staurolite is distinguished by its hardness and its propensity for growing two "twin" crystals at the same time. The twins are commonly oriented perpendicularly to each other, forming a cross. Some collectors call these "fairy crosses," and they are considered good-luck charms. Other twins intersect at about 60-degree angles.

Staurolite crystals are medium to dark in color, usually reddish brown, tan, brown, or black. Prismatic crystals have a vitreous luster and are hexagonal or diamond shaped in cross section. Crystals with well-developed faces have a glassy luster and are hexagonal or diamond shaped in cross section.

Epidote

Class: Silicate Chemical formula: $Ca_2(Al,Fe)_3(SiO_4)_3(OH)$

Epidote is a common mineral with a characteristic pistachio green color. It is usually a secondary mineral, grown during metamorphism or alteration of a rock.

KEY FACTS

The green colors of epidote are distinctive. This mineral has one direction of perfect cleavage—a characteristic distinguishing it from green amphiboles, which have two cleavage planes.

+ **hardness:** 6–7

+ **streak:** Colorless to gray

+ **locations:** Found in metamorphic rocks, pegmatites, and small cavities in basalt

Epidote is a common mineral found in rocks that have undergone metamorphism or alteration, often by the heat of a nearby magma chamber. Epidote forms well-developed columnar prisms or thick, tabular crystals with grooved faces. It also forms thin crusts and seams, filling vesicles (small cavities) and fractures in basalts. Epidote has a characteristic green color, often pistachio green, yellow-green, or greenish black. It has a vitreous luster, and it is transparent to translucent. Epidote crystals are pleochroic, meaning that different colors appear through the crystal prism as the prism is rotated.

Beryl

Class: Silicate Chemical formula: $Be_3Al_2Si_6O_{18}$

Beryl is best known for its gemstone varieties emerald and aquamarine. It is an aluminosilicate mineral with beryllium in its crystal structure. It is most often found in granitic pegmatites.

KEY FACTS

The beautiful colors of highly prized gem-quality beryl include red, pink, green, and blue. The crystals form hexagonal prisms, which add to the mineral's popular appeal.

+ hardness: 7.5–8

+ streak: White

+ locations:
Although usually found in granitic pegmatites, beryl also occurs in some metamorphic rocks.

Beryl is often recognized by its hexagonal crystals. Beryls form beautiful prisms in granitic pegmatites. The beryl crystal structure contains relatively large spaces that can accommodate additional atoms of different elements, especially manganese, iron, and chromium, resulting in a wide variety of colors. Beryl with iron in the beryllium site forms the gem aquamarine. Manganese in the aluminum site forms the pink gem morganite. Chromium beryls are emeralds. Beryl crystals have a vitreous to greasy luster and are harder than quartz. In addition to granitic pegmatites, beryls can also be found in some metamorphic rocks.

Tourmaline

Class: Silicate Chemical formula: $Na(Mg,Fe)_3Al_6(BO_3)_3(Si_6O_{18})(OH,F)_4$

Tourmaline forms beautiful columnar crystals with rounded triangular cross sections. Its colors include black, brown, blue, pink, and green, sometimes with several colors in one crystal.

KEY FACTS

Distinguishing characteristics of tourmaline include well-formed columnar crystals with pyramidal faces on one end. Some are multicolored, including a type popularly called "watermelon tourmaline."

+ **hardness: 7–7.5**

+ **streak: White**

+ **locations: Found in granites, pegmatites, quartz veins, and some metamorphic rocks**

Tourmaline crystals have rough, vertical striations and rounded to triangular cross sections. Crystals often have pyramidal faces on one end. Tourmaline is most commonly the black, vitreous variety known as schorl. A popular gem called "watermelon tourmaline" displays concentrically zoned colors with red or pink inside and green outside, formed by chemical changes during crystal growth. Red or pink tourmalines contain manganese; green tourmalines result from iron. The largest crystals are found in pegmatites, though tourmaline is also found in schists and gneisses. Gem-quality tourmalines are collected in Maine, Connecticut, and California.

Pyroxene

Class: Silicate Chemical formula: $XY(Si,Al)_2O_6$

Pyroxenes are a group of important rock-forming minerals found in igneous and metamorphic rocks. The most common pyroxene mineral is augite, a component of some volcanic rocks.

KEY FACTS

These silicates are similar to amphiboles but have cleavage planes intersecting at right angles.

+ hardness: 5–6.5

+ streak: Variable; augite light green to colorless.

+ locations: Mafic and ultramafic igneous and metamorphic rocks; forms large crystals in porphyritic volcanic rocks; also occurs as granular masses

The pyroxene group contains various minerals defined by crystal structures and chemical compositions, including aegirine, augite, diopside, jadeite, and enstatite. These are high-temperature minerals similar to amphiboles but without water content. Pyroxenes form prismatic crystals with vitreous luster, in elongate and short varieties. Most pyroxenes require a hand lens to identify. They differ from amphiboles by their cleavage angles: pyroxene intersecting at right angles; amphibole intersecting in wedge or diamond shapes. Earth's upper mantle is made of olivine and pyroxene. Augite is found in volcanic dikes.

Amphibole

Class: Silicate Chemical formula: $XY_2Z_5(Si,Al,Ti)_8O_{22}(OH,F)_2$

Amphiboles are a group of important rock-forming minerals. They are typically dark in color and are distinguished by their diamond- or wedge-shaped cleavage intersections.

KEY FACTS

Elongate crystals are generally flatter than pyroxenes, with oblique, perfect cleavage planes that intersect at approximately 120 degrees.

+ **hardness:** 5–6

+ **streak:** Variable; usually white or colorless

+ **locations:** Found in many igneous and metamorphic rocks including granite, basalt, gneisses, schists, and amphibolite

Like pyroxenes, amphiboles are minerals rich in iron and magnesium. Unlike pyroxenes, amphiboles are hydrous, meaning they have water in their structure. Geologists sometimes call amphiboles "garbage-can minerals" because many elements fit into the crystal structure. The most common amphibole is hornblende, a constituent of granitic rocks. Others include tremolite and actinolite in schists and metamorphosed impure limestones. Amphiboles form elongate, prismatic crystals, sometimes in radiating aggregates, typically gray (tremolite) to dark green (actinolite). Darker colors indicate increasing iron content.

Hornblende

Class: Silicate Chemical formula: $(Ca,Na,K)_{2\text{-}3}(Mg,Fe,Al)_5(Si,Al)_8O_{22}(OH)_2$

Hornblende is the name for various iron- and magnesium-rich amphiboles found in many igneous and metamorphic rocks.

KEY FACTS

A distinguishing characteristic of hornblende is its wedge-shaped amphibole cleavage. Its most common colors are black and dark brown.

+ hardness: 5–6

+ streak: Colorless to white, gray, or brown

+ locations: Found in many igneous and metamorphic rocks; large, well-formed crystals occur in granitic pegmatites

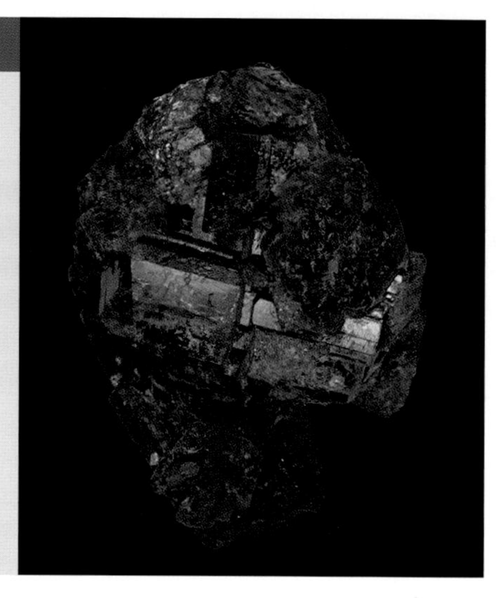

Hornblende is usually black or dark brown, but can be dark green. It is distinguished from other dark minerals such as biotite, tourmaline, and pyroxenes by its wedge-shaped cleavage angles. Cleavage angles are usually identified under a hand lens. Well-formed crystals are commonly short columnar prisms with six-sided cross sections and a vitreous to dull luster. Hornblende crystals can have a bladed or even fibrous habit and form massive aggregates. Hornblende is a common constituent of many rocks including granite, syenite, diorite, gabbro, basalt, schist, gneiss, and amphibolite.

Talc

Class: Silicate Chemical formula: $Mg_3Si_4O_{10}(OH)_2$

Talc is a soft, sheetlike mineral. It is the main component of soapstone and is also used in lubricants, cosmetics, and in industrial applications as a heat-resistant material.

KEY FACTS

Its greasy touch and low hardness easily identify talc. It is soft enough to be scratched by a fingernail, although talc schist is used as a building material.

+ **hardness:** 1
+ **streak:** White
+ **locations:** Metamorphic rocks including schists, marbles, and metaperidotites; associated with serpentine, pyroxene, and olivine

Talc is an important commercial and industrial material, used for its heat-resistant and lubricant properties. It is found in metamorphic terrains associated with ocean floor and ultramafic rocks such as marbles and peridotites. The common building material soapstone is a talc schist. Talc forms white to light green aggregate masses with a distinctive greasy touch and low hardness. It rarely forms individual crystals, but can replace crystals of other minerals during alteration, and can take on their shape, known as pseudomorphs. Talc is found, among other places, in the Appalachians, California, and Texas.

Muscovite

Class: Silicate Chemical formula: $KAl_3Si_3O_{10}(OH,F)_2$

Muscovite is one of the most common micas, a group of sheetlike minerals called phyllosilicates. It is the clear, light brown, or gray mica found in schists, gneisses, and sometimes in granites.

KEY FACTS

This widespread form of mica has thin, flexible, transparent cleavage sheets and is generally light gray or silvery. It has important uses in electronics.

+ hardness: 2–3

+ streak: Colorless or white

+ locations: Granitic rocks, pegmatites, schists, and gneisses; large crystals found in various areas of the U.S.

Muscovite is a common mica, easily identified by its flaky habit and perfect single cleavage that creates easy-to-peel, thin, flexible sheets. Muscovite is distinguished from biotite by its light colors, usually clear, white, gray, or silvery. It typically forms larger crystal masses than biotite and is more common. Large sheets of muscovite are mined from granitic pegmatites for use in the electronics industry. The name refers to the Muscovy region in Russia where large sheets of muscovite were once used as window glass. Large crystals are common in North Carolina, New England, Colorado, and South Dakota.

Biotite

Class: Silicate Chemical formula: $K(Mg,Fe)_3(Al,Fe)Si_3O_{18}(OH,F)_2$

Biotite, along with muscovite, is part of the mica group of phyllosilicates, or sheetlike minerals. It is a platy black mineral found in some granites and schists.

KEY FACTS

Black or deep brown plates (which are tabular crystals) have a distinctive flaky cleavage. Sometimes a gold tint to the plate gives this mineral a resemblance to pyrite.

- **hardness: 2–3**
- **streak: Colorless or white**
- **locations: Commonly a component of schists, but also can be found in biotite granites**

Biotite is a black or deep brown mica. Gold-tinted biotite can be distinguished from pyrite by its platy cleavage and low hardness. Large biotite "books" can be found in granitic pegmatites. Well-formed crystals can have a hexagonal cross section, and they can easily peel into thin, flexible sheets. Most biotite, however, is found as small, embedded dark-colored grains.

Biotite is an important mineral in geological research, used for analyzing the temperature histories of metamorphic rocks and, in some cases, for determining the approximate ages of igneous rocks.

Serpentine

Class: Silicate Chemical formula: $Mg_3Si_2O_5(OH)_4$

Serpentines are a group of minerals that form by the alteration of olivine and pyroxene. Serpentines are associated with rocks that originated in the oceanic crust and upper mantle.

KEY FACTS

This silicate is commonly dark green with a greasy to waxy luster and feel; compact and massive. The name suggests its resemblance to snake skin.

+ hardness: 2–5

+ streak: White

+ locations: Zones of altered rocks, associated with other magnesium-rich minerals. It is commonly found along California's coast ranges.

Serpentine minerals have three main varieties: antigorite, chrysotile, and lizardite. Antigorite is most common, occurring in compact masses that are typically dark green with a greasy luster. Lizardite is fine grained and scaly, and is found associated with altered marbles. Chrysotile is a type of asbestos with a fine fibrous form and very low hardness. Masses of serpentine make up rocks called serpentinites; these can have irregular green patterns and scaly surfaces resembling a snake skin, hence the name "serpent rock." Rocks of serpentine masses and serpentine breccia are used as decorative building stones.

Kaolinite

Class: Silicate Chemical formula: $Al_2Si_2O_5(OH)_4$

Kaolinite is the main component of clay. It forms compact masses that can be layered. It is either a primary mineral, formed in place, or it is the weathering product of minerals such as feldspar.

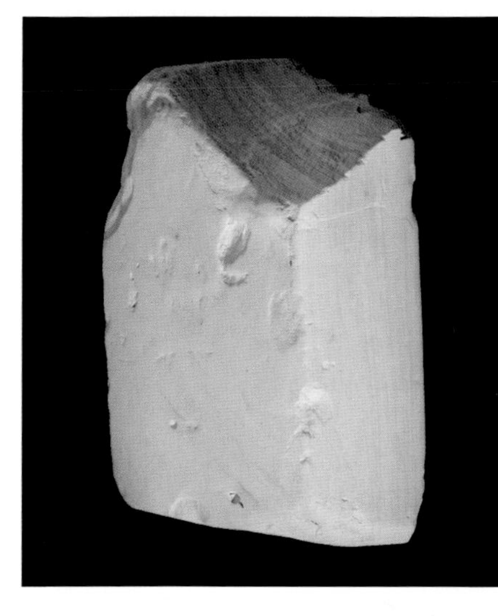

KEY FACTS

Compact, flourlike masses are usually white or earthy red, and they have an earthy odor. This clay has a great variety of uses, ranging from cosmetics to china to construction bricks.

- **hardness:** 2
- **streak:** White
- **locations:** Found in clay beds and in place of altered feldspars in granites and pegmatites

Kaolinite is a type of clay found in areas of chemical weathering. It forms microscopic platy crystals that make up compact, earthy masses. It has a dull luster, is relatively soft, has a distinct earthy odor, powdery feel, and is usually white. Illite and montmorillonite are other common clay minerals in soils. It is difficult to distinguish these without laboratory equipment. Kaolinite is one of the most important nonmetallic minerals, used in many commercial and industrial applications including cosmetics, paints, adhesives, and glossy paper. Kaolinite clay is also used to make bricks, tiles, fine pottery, and china.

Calcite

Class: Carbonates Chemical formula: $CaCO_3$

Calcite is the primary mineral component of limestone and marble. It is the stable form of calcium carbonate at surface temperatures and pressures, and often has a biological origin.

KEY FACTS

Its "fizzing" in dilute acids (for example, hydrochloric acid) is a characteristic feature of calcite. Its hardness is also diagnostic.

- **hardness: 3**
- **streak: White**
- **location: Component of carbonate rocks including limestone, travertine, and marble; also forms thin veins and the linings of small cavities**

Two common forms of calcium carbonate are calcite and aragonite. Calcite is the stable form at the Earth's surface. It is typically clear or white, but can be yellow or gray, with a pearly or vitreous luster. It is found in many different crystal habits and shapes: rhombohedrons, prisms, granular masses, and microcrystalline masses. Crystals display three perfect cleavage planes. Calcite differs from other white or clear minerals by its hardness—unlike quartz, it can be scratched by a knife—and its fizzing reaction in dilute acids. Dolomite, on the other hand, will fizz only under a concentrated hydrochloric acid solution.

Dolomite

Class: Carbonates Chemical formula: $CaMg(Co_3)_2$

Dolomite is an important rock-forming mineral in sedimentary and some metamorphic rocks. It is the primary component of the rock known as dolomite or dolostone.

KEY FACTS

Although it is similar to calcite, dolomite reacts only to concentrated hydrochloric acid.

+ **hardness:** 3.5–4

+ **streak:** White

+ **location:** Besides its occurrence in the rock dolomite, the mineral also occurs with calcite in limestone and some marbles. It can be found in hydrothermal veins and small cavities where it forms crystals.

Dolomite is similar to calcite but with magnesium in its structure as well as calcium. Dolomite crystals are commonly rhombohedrons that are curved or slightly saddle shaped; like calcite, it forms granular and microgranular masses. Dolomite is typically white, gray, brown, or yellow. Its hardness, luster, and cleavage are similar to calcite. Calcite, however, will fizz under a diluted hydrochloric acid solution, whereas dolomite requires concentrated acid or a scratched or powdered surface to cause a fizzing reaction.

Malachite

Class: Carbonates Chemical formula: $Cu_2CO_3(OH)_2$

Malachite is a beautiful green, copper carbonate mineral. It is often associated with copper ore deposits and coexists with azurite, a less common copper carbonate that is brilliant blue.

KEY FACTS

This attractive carbonate is an opaque, green, banded mineral. It can be cut and polished to a high luster and is often used in jewelry.

+ **hardness:** 3.5–4
+ **streak:** Green
+ **location:** Occurs in association with copper ores, especially in the copper mining region of the American Southwest. It is also found in limestones.

Malachite is an opaque, striking green mineral, often displaying color bands of different shades of green. It is typically found in massive aggregates with a globular or bubbly surface (geologists call this botryoidal or mammillary habit), but can also form thin coatings or stalactites in caverns, or veins/veinlets along fractures in limestone. Well-developed crystals are rarely true malachite crystals; instead, they are pseudomorphs that form by replacing crystals of azurite. Malachite masses take a beautiful high polish. Pure specimens are cut and polished into various forms and collected as semiprecious or decorative stones.

Gypsum & Anhydrite

Class: Sulfates Chemical formula: $CaSO_4 \cdot 2H_2O$

Gypsum and anhydrite are calcium sulfate minerals, gypsum as the hydrous form and anhydrite the form without water ($CaSO_4$). They are typically found in evaporite deposits.

KEY FACTS

Gypsum forms flattened, glassy crystals that are sometimes clear; anhydrite is usually found as veins or larger masses.

+ **hardness:** 1.5–2 (gypsum); 3–3.5 (anhydrite)
+ **streak:** White
+ **location:** Found in evaporite deposits as massive sedimentary beds; also in association with clay beds. A vast example is New Mexico's White Sands.

Gypsum and anhydrite occur in deposits formed when lakes and shallow seas evaporate, and from waters circulating through sandstones and clays. Crystals are colorless to white; can be tabular or diamond shaped/rhombic. Radiating blades of gypsum coated in sand are called desert roses. Glassy gypsum crystals are called selenite. Massive gypsum beds are known as alabaster. Gypsum is mined to produce cement, Sheetrock, and fertilizer. It makes up the dramatic dunes of White Sands National Monument in New Mexico. Anhydrite is harder and has three good cleavage planes that create cubic-like fragments.

Galena

Class: Sulfides Chemical formula: PbS

Galena is a gray metallic mineral typically found in hydrothermal ore deposits. Galena is made of atoms of lead and sulfur packed together in a cubic structure and is an important lead ore.

KEY FACTS

This important ore is a distinctive lead-gray color with a metallic luster, and a cubic or octahedral shape. A special characteristic is how heavy it feels in the hand.

- **hardness: 2.5**
- **streak: Gray**
- **location: Found in hydrothermal deposits, often in association with sphalerite, pyrite, and chalcopyrite**

Galena is a major lead ore and a source of silver, bismuth, and thallium. It commonly forms perfect cubes or octahedrons. Galena is opaque gray with a shiny metallic luster when fresh. Specimens easily tarnish with exposure to air, giving some samples a dull luster. Hardness is a key identifying characteristic—it can be scratched with a fingernail—yet it has a high specific gravity, so it feels heavy in the hand. The distinctive gray streak is also diagnostic. Galena is common in silver and lead mining areas of the U.S. and Canada. It is the state mineral of Missouri and Wisconsin.

Chalcopyrite

Class: Sulfides Chemical formula: $CuFeS_2$

Chalcopyrite, a copper iron sulfide, is an important copper ore. It is a brassy and golden mineral found in hydrothermal ore deposits and is a common specimen in rock and mineral collections.

KEY FACTS

Brassy color differs from the bright golden hue of gold and pyrite. Specimens are easily tarnished. Chalcopyrite is softer than pyrite and harder than gold.

+ hardness: 3.5

+ streak: Green-black

+ location: Ore deposits, primarily in hydrothermal veins. An important location is in the greenstone belt of Ontario.

Chalcopyrite is distinguished by its greenish brassy color, distinct from the metallic gold of both pyrite and gold. Its hardness is also distinctive. It is softer than pyrite, whereas gold is softer than both chalcopyrite and pyrite, and is more malleable. Well-formed chalcopyrite crystals are rare and typically have a tetrahedral shape, which distinguishes them from the cubic form of pyrite. Chalcopyrite occurs in sulfide ore deposits formed by deposition of copper during the circulation of medium-temperature fluids in volcanic environments. It occurs as mineral veins, granular masses, and nodules.

Pyrite

Class: Sulfides Chemical formula: FeS_2

Pyrite, commonly known as fool's gold, is made of iron sulfide. It forms distinctive golden, metallic cubic crystals that are popular with collectors.

KEY FACTS

Pyrite's hardness is diagnostic—much harder (but lighter) than gold, and it can scratch glass. Its streak is also a helpful identifier.

- **hardness: 6–6.5**
- **streak: Greenish black**
- **location: Hydrothermal veins and other sites in sedimentary, metamorphic, and igneous environments. Small crystals can be found in shale.**

Pyrite is known as fool's gold because of its gold color and high metallic luster. Unlike true gold, however, pyrite is relatively hard—similar in hardness to quartz—and is commonly found in the form of distinctive crystal cubes or crystals with 12 sides in the shape of pentagons. It is also significantly lighter than gold. Crystal faces of pyrite commonly display striations. Another diagnostic feature of pyrite is that it gives off a sulfur smell when broken or hammered. Pyrite is the most common sulfide mineral and is used to produce sulfuric acid. It is sometimes found as a replacement mineral in fossils.

Halite

Class: Halides Chemical formula: NaCl

Halite is the naturally occurring form of table salt and is also known as rock salt. It is found in evaporite deposits and can form thick beds in sedimentary sequences.

KEY FACTS

Distinguished by its salty taste and solubility in water, halite has long been familiar as table salt; however, current grocery store salt is primarily synthetic.

+ **hardness: 2.5**
+ **streak: White**
+ **location: Evaporite deposits and in salt crusts along shorelines in arid environments**

Halite is used as road salt and is important in the chemical industry—for example, to produce hydrochloric acid. Salt crystals are cubic, and are typically clear, white, or sometimes bluish or with blue spots. They have a vitreous or greasy luster. Halite precipitates out of seawater when the water evaporates to less than 10 percent of its original volume. Because of halite's low density, salt beds can rise under pressure through weaknesses in overlying rocks. These disruptions of sedimentary strata create interesting structures in the deserts of the U.S. Southwest.

Fluorite

Class: Halides Chemical formula: CaF_2

Fluorite is a colorful mineral that forms cubic or octahedral crystals. Fluorite fluoresces under ultraviolet light, and the phenomenon is named after this property of fluorite.

KEY FACTS

Transparent to translucent crystals of fluorite come in various colors, including pale purple, pink, green, and yellow. Its fluorescence under ultraviolet light is a famous characteristic.

+ **hardness:** 4

+ **streak:** White

+ **location:** Occurs in hydrothermal veins; also in cavities of granitic pegmatites

Fluorite is usually found as well-formed crystals with cubic and octahedral faces. It can be colorless, but is more often colorful, including dark blue, violet, pink, green, and yellow varieties. The colors are influenced by concentrations of rare earth elements in the crystal structure. Crystals have a vitreous luster. Hardness is a good diagnostic tool. It is softer than quartz but harder than calcite. Some specimens show cleavage planes intersecting at 60 degrees. Fluorite forms in medium- and low-temperature hydrothermal deposits, in crystal pockets of granites and granitic pegmatites, and as a component of some metamorphic rocks.

Silver

Class: Native elements Chemical formula: Ag

Silver is a rare native element that occurs in hydrothermal deposits and in association with other ores. Silver is distinctively malleable and is an important precious metal.

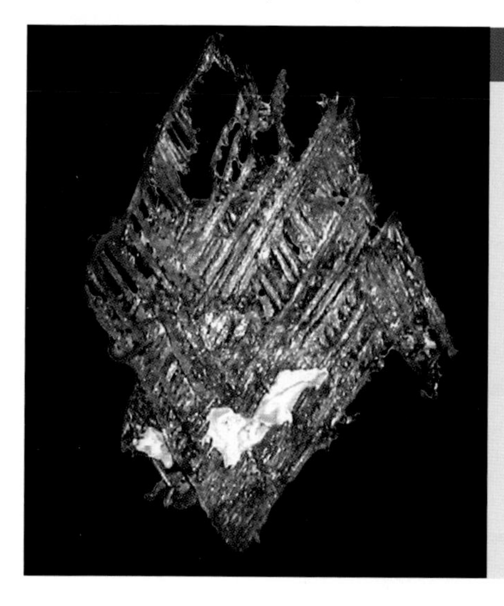

KEY FACTS

This prized metal has a silver-white color and a high metallic luster when it is fresh; however, as is well known, its surface oxidizes or tarnishes easily to dark gray and black.

+ hardness: 2.5

+ streak: Silver-white

+ location: Hydrothermal veins or by alteration of other minerals; also placer deposits

Native silver is rare, found in hydrothermal deposits and alteration zones (where high-temperature fluids have either deposited material or changed existing rocks). It occurs in thin plates, as well as in granular habits and skeletal and wirelike, branching forms. It is distinguished by its opaque, silver-white color and metallic luster that easily tarnishes. It is distinctly malleable and is relatively soft. Most silver is produced by refining other compounds instead of mined in its pure form. It is used as currency and in jewelry, and is important industrially for its conductivity and malleability.

Gold

Class: Native elements Chemical formula: Au

The precious metal gold is used for currency, jewelry, electronics, and more. It occurs in hydrothermal deposits and in quartz veins, but is mostly found as grains in river and beach sands.

KEY FACTS

Known for its distinctive color and metallic luster, gold resists oxidation. Most gold is discovered in accumulations of particles in river and beach sands.

+ hardness: 2.5–3

+ streak: Golden

+ location: Hydrothermal deposits, quartz veins in large granitic bodies, and secondary weathering deposits

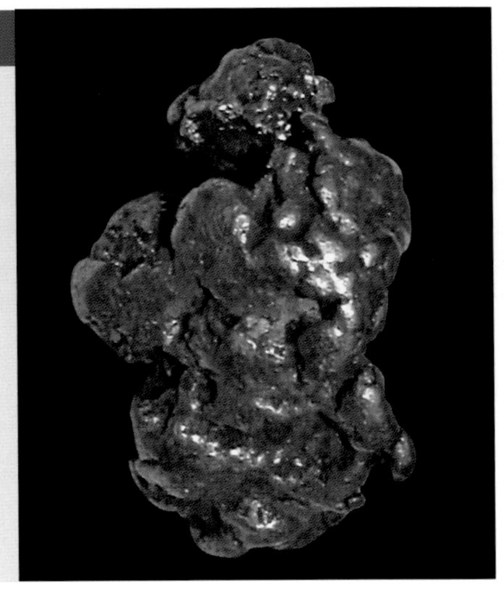

Gold rarely forms distinct crystals; instead, it is found as golden yellow platelets, branching, wiry forms, and as inclusions in quartz. Gold is soft, but has a very high density. It is malleable, and it maintains its metallic luster and color without oxidation. Gold differs from pyrite by its hardness (gold is softer) and by a lack of sulfur smell. Flecks of yellowish mica can resemble gold, but mica has a vitreous, not metallic, luster and is brittle. Gold is extremely unreactive and survives chemical weathering. Most gold is found as particles in gravels and sands that have weathered from gold-bearing veins.

Copper

Class: Native elements Chemical formula: Cu

Native copper occurs as irregular nodules, veins, and wirelike or platelike forms. It is an important industrial metal because of its high thermal and electrical conductivity.

KEY FACTS

The original copper-red color is often oxidized to black, blue, and green. Native copper is usually found in irregular masses, often with strange shapes.

+ hardness: 2.5–3

+ streak: Copper-red

+ location: Hydrothermal veins; also in rare lava flows including along the mid-continent rift in northern Michigan

Most industrial copper is extracted from other minerals such as chalcopyrite, but native copper can be found. Copper rarely exists as well-formed crystals and is more often found as irregular masses, sometimes with branching or wirelike shapes, and as the matrix surrounding volcanic clasts. A key to identifying copper is its copper-red streak and low hardness—it can be scratched by a knife. Copper specimens commonly display thin black, blue, or green oxidation coatings. One of the world's largest deposits of native copper is in Michigan's Upper Peninsula, where it is associated with a thick series of lava flows.

Diamond

Class: Native elements Chemical formula: C

Diamond, the hardest mineral, consists of pure carbon, and in nature, forms only at high pressures and temperatures deep in the Earth. Diamonds come to the Earth's surface via volcanic vents.

KEY FACTS

A diamond's famed hardness is diagnostic—nothing will scratch it. The various colors of diamonds are results of impurities such as the elements boron and nitrogen.

+ hardness: 10

+ streak: None

+ location: Rare; found in association with ultramafic (iron- and magnesium-rich) igneous rocks in continental shields

Diamonds are a high-pressure crystalline form of carbon with individual atoms packed together by strong chemical bonds. Graphite is also pure carbon, but is a low-pressure mineral in the form of carbon sheets with weak bonds. Diamonds are unstable at the Earth's surface temperatures and pressures, but they cannot convert to graphite on human timescales. Small fragments of quartz can look like diamonds but are distinguished by testing hardness. Diamond crystals are usually octahedrons or cubic forms. Impurities give diamonds different colors: Boron, for example, creates a blue tint; nitrogen casts a yellow hue.

Magnetite

Class: Oxides Chemical formula: Fe_3O_4

Magnetite is a form of iron oxide found in some metamorphic and igneous rocks as well as secondary deposits in black sands. True to its name, its magnetism is diagnostic.

KEY FACTS

This magnetic oxide is black, with a semi-metallic luster. Black sands often found along rivers and on ocean beaches are grains of magnetite that have eroded out of rocks.

+ **hardness:** 5.5–6
+ **streak:** Dark gray to black
+ **location:** Common accessory mineral in a variety of metamorphic and igneous rocks

Magnetite is a major iron ore, found as octahedral crystals, granular aggregates, and in tiny veins. It is dark gray to black with metallic luster. Its black streak and magnetism are distinguishing characteristics. Magnetite is important to geologists because it records information about the Earth's magnetic field when it crystallizes. Analyzing magnetite-bearing rocks of different ages allows geologists to understand movement of the Earth's tectonic plates. Large crystals of magnetite occur in pegmatites and in high-temperature veins.

Hematite

Class: Oxides Chemical formula: Fe_2O_3

Hematite is the most common form of iron oxide and an important iron ore. It occurs in various forms including black platy or tabular crystals and irregular masses with a brick red hue.

KEY FACTS

A red streak is diagnostic for this major source of iron, found as thick sedimentary beds in some formations. Crystals have a metallic luster, but masses are typically dull.

+ hardness: 5–6

+ streak: Bright to earthy red

+ location: A common component of many different rocks; large deposits are found in banded iron formations.

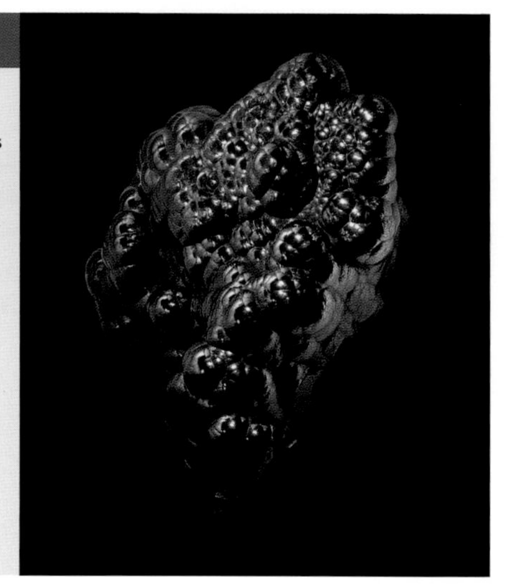

Hematite varies in form and color, appearing steely gray, black, brown, orange, or red. The red streak is a good method to differentiate it from similar minerals such as magnetite and ilmenite. Hematite can form circular aggregates of platy crystals known as iron roses. It also forms tabular and hexagonal crystals, and irregular, rounded masses. Clays with high hematite content are called red ochre, which is used as a pigment. Hematite forms as a deposit in quartz veins, as well as in metamorphic rocks. It is the primary component of banded iron formations.

Corundum

Class: Oxides Chemical formula: Al_2O_3

Corundum is an aluminum oxide known for its high hardness and beautiful gemstone varieties, including pink and red rubies, and blue, green, and yellow sapphires.

KEY FACTS

The hardness of corundum is diagnostic; only diamonds are harder. This mineral forms tabular and prismatic crystals. Impurities provide colored gems such as rubies and sapphires.

- **hardness: 9**
- **streak: White**
- **location: Pegmatites, gneisses, schists, and marbles; also in association with ultramafic igneous rocks**

Corundum is an aluminum oxide mineral with a close-packed crystal structure and strong bonds between atoms, giving it high hardness and density. Corundum of pure aluminum oxide is clear, white, or gray. Impurities lead to various colors. Chromium creates pink or red rubies. Iron and titanium create blue sapphires; iron alone creates yellow sapphires. Corundum forms during alteration of volcanic and ultramafic igneous rocks and during metamorphism of aluminum-rich sedimentary and igneous rocks. Corundum is used in industry as an abrasive. Corundum and quartz never coexist in the same formation.

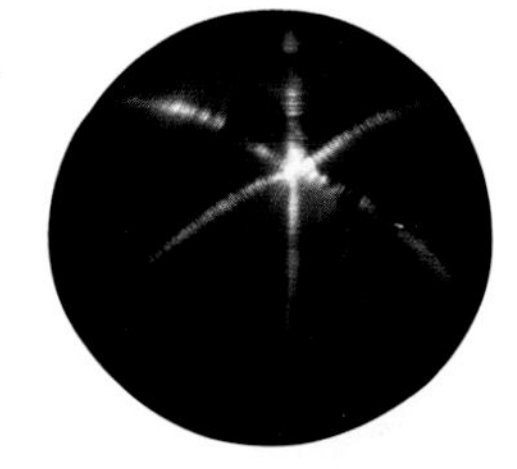

Granite

Type: Igneous

Granite is one of the Earth's most common rocks and an important building block of the continents. It is a light-colored intrusive igneous rock with a distinct granular texture.

KEY FACTS

A relatively homogeneous rock with interlocking minerals, granite can be grayish or light pink to brick red; exposed by erosion.

+ grain size: Medium to coarse

+ texture: Crystalline; uniform grain size to porphyritic

+ composition: Quartz + alkali feldspar + plagioclase feldspar ± hornblende ± biotite ± muscovite

Granite is an intrusive igneous rock containing at least 20 percent quartz and two types of feldspar—plagioclase and alkali feldspar in relatively equal amounts. It can be difficult to distinguish between the feldspars in the field, or to determine the amount of quartz; thus, many medium-grained, light-colored igneous rocks are called granite, even if they are technically something else.

Granites form from felsic magma that slowly cools below the Earth's surface. Multiple intrusions of granitic magma build up large rock bodies called batholiths, which form in the cores of mountain belts and are exposed by erosion.

Porphyritic Granite

Type: Igneous

A porphyritic granite has mineral grains of two different sizes. Typically these are large feldspar grains, often with well-formed crystal faces, surrounded by smaller grains of other minerals.

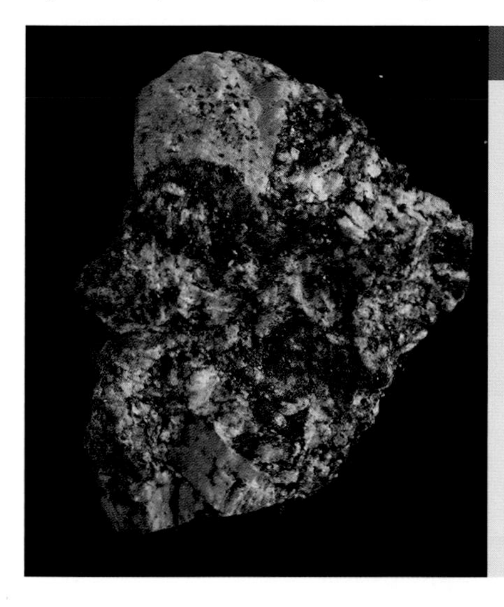

KEY FACTS

This form of granite commonly has large pink or red alkali feldspar grains. It is often the material of monuments.

+ grain size: Medium to coarse

+ texture: Crystalline; magmatic alignments of large crystals are possible.

+ composition: Quartz + alkali feldspar + plagioclase feldspar ± hornblende ± biotite ± muscovite

The term "porphyritic" describes igneous rocks that have minerals with two grain sizes. The larger grains are called phenocrysts, from the Greek *phaino,* meaning "visible," and *cryst,* meaning "crystal." Porphyritic granites often have pink or red alkali feldspar phenocrysts surrounded by smaller grains of plagioclase, quartz, biotite, and/or hornblende. These phenocrysts form when the alkali feldspar is the first mineral to crystallize out of the magma. The feldspars grow before the other minerals crystallize and fill the remaining space. Porphyritic granites are used for countertops, wall claddings, and monuments.

Rapakivi Granite

Type: Igneous

Rapakivi granites are granites with a striking texture of large, rounded feldspar grains with different colored rims. The name "rapakivi" is Finnish and means "crumbly stone."

KEY FACTS

This granitic rock is used as a building and decorative interior stone.

+ grain size: Medium to coarse

+ texture: Crystalline, porphyritic

+ composition: Quartz + alkali feldspar + plagioclase feldspar ± hornblende ± biotite ± muscovite

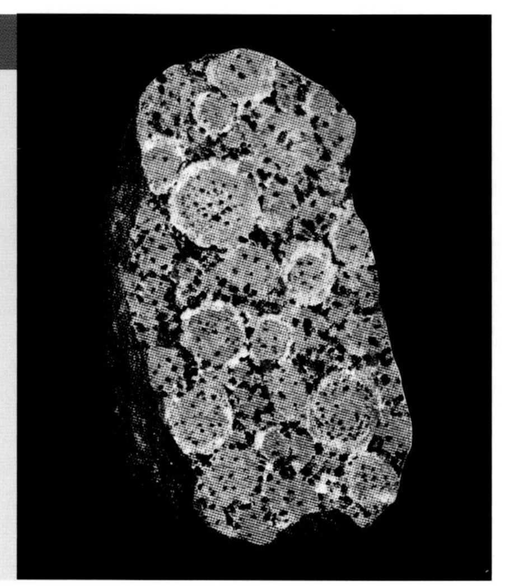

Rapakivi granites are common decorative stones with large pink or brown alkali feldspar grains surrounded by gray or white plagioclase rims. The large grains, called phenocrysts or megacrysts, typically have a rounded shape. Rapakivi granites are found on all continents, usually in the older continental cores. The building stone known as Baltic brown is a rapakivi granite from Finland that is about 1.6 billion years old. It is a popular stone used for kitchen countertops and flooring. The groundmass of this granitic rock is a dark color because it is rich in biotite and hornblende.

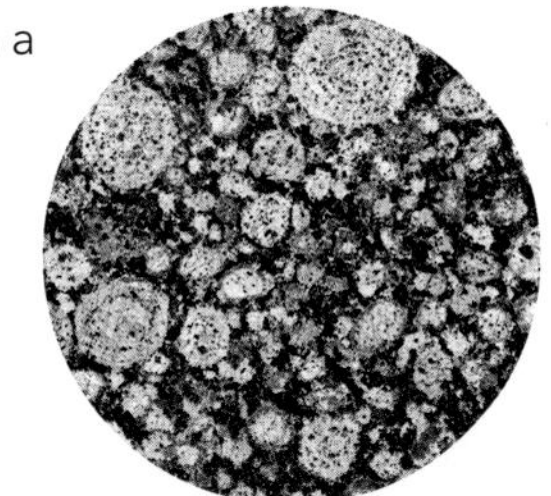

Aplite

Type: Igneous

Aplites are light-colored, fine-grained intrusive rocks that commonly form dikes. Most are found associated with crosscutting large granitic bodies and have a composition similar to granite.

KEY FACTS

These rocks have an even, fine grain size, creating a sugary texture.

+ **grain size:** Fine to medium

+ **texture:** Crystalline; even grain size, no fabric

+ **composition:** Quartz + alkali feldspar + plagioclase ± mica

Cutting across many exposures of granitic rocks are relatively narrow dikes that are conspicuously light in color and fine in grain size. These are called aplite dikes and they typically have bulk compositions that are similar to granite. Although the grains are small, quartz, feldspar, and sometimes muscovite are usually visible. Aplites are thought to be formed from the last remnants of the magma that created the surrounding granite. This magma must have cooled quickly to produce the fine texture. Aplites are often more resistant to weathering than the surrounding rock, and can project out as ridges.

Pegmatite

Type: Igneous

A pegmatite is an intrusive igneous body with very coarse-grained, interlocking crystals. Many pegmatites are granitic. Pegmatites are sites for finding rare mineral specimens.

KEY FACTS

Crosscutting dikes, veins, or irregular intrusions with very large grains are features of pegmatite.

+ grain size: Very coarse

+ texture: Crystalline; grain size and mineral distribution can be heterogeneous.

+ composition: Variable; most pegmatites are granitic.

Pegmatites are intrusive bodies of rock notable for very coarse grain size. The crystals vary from a few inches to several yards across. Pegmatite intrusions can be dikes, veins, and lenses that cut across preexistent rock. Geologists differ about how pegmatites form; most agree that they form from the last remaining melt in a magma chamber, which is saturated in water and other fluids. This melt must cool very slowly to grow such large crystals. Its concentration of rare elements leads to crystallization of exotic minerals, including many gems. Pegmatites are important sources for mining feldspar and mica.

Granodiorite

Type: Igneous

Granodiorite is an intrusive igneous rock that resembles granite but contains less alkali feldspar and more plagioclase than a true granite. It is a common building and decorative stone.

KEY FACTS

Usually light to medium gray overall, granodiorite has interlocking white, gray, and black mineral grains.

+ **grain size:** Medium to fine
+ **texture:** Crystalline
+ **composition:** Quartz + plagioclase + alkali feldspar ± hornblende ± biotite

Granodiorite has an intermediate composition between felsic granite and mafic gabbro. It has an overall light- to medium-gray appearance, dominated by white or gray grains of plagioclase and quartz, plus black grains of biotite and hornblende. Granodiorite is often lumped with granite and described as a granitic rock or granitoid. It is a common building and decorative stone, used for countertops, building facades, and structural blocks. The rocks of the Sierra Nevada are primarily granites and granodiorites, with granodiorite making up many iconic formations in Yosemite National Park.

Syenite

Type: Igneous

Syenite is an uncommon intrusive igneous rock similar to granite but with little or no quartz and abundant alkali feldspar. Nepheline syenites are syenites that contain nepheline and no quartz.

KEY FACTS

Syenite is dominated by alkali feldspar grains.

+ grain size: Medium to coarse

+ texture: Crystalline, sometimes porphyritic

+ composition: Alkali feldspar ± quartz or feldspathoid (nepheline) ± plagioclase ± amphibole or pyroxene ± biotite

Syenite is an intrusive igneous rock that is predominantly alkali feldspar. Quartz is either absent or present in small amounts. If the rock contains nepheline, which is never present with quartz, it is known as a nepheline syenite. Syenites are formed in the thick continental crust from magma derived from partially melting granitic rocks, and are associated with failed continental rift valleys. Syenites are mined for use in the glass and ceramics industries to lower the melting temperature of a glass or ceramic mixture. They are also used as decorative stones, especially a blue variety that gains its color from the blue mineral sodalite.

Diorite

Type: Igneous

Diorite is a medium- to dark-gray intrusive igneous rock made of plagioclase feldspar and a dark mineral, usually hornblende. It is commonly found in association with granite and granodiorite.

KEY FACTS

Diorite has a salt-and-pepper appearance with mixed light and dark grains.

- **grain size: Medium to coarse**
- **texture: Crystalline; can be porphyritic**
- **composition: Plagioclase + hornblende ± biotite**

Diorite is an intrusive igneous rock made up of predominantly plagioclase feldspar plus an iron and magnesium mineral component, typically hornblende, sometimes with biotite. It contains less quartz than granodiorite and less quartz and alkali feldspar than granite. Because of its mineral assemblage, diorite is medium to dark gray in overall color, with a salt-and-pepper appearance created by the light plagioclase grains and black hornblende grains. Inclusions of diorite are common in some granitic bodies. Diorite is used as a structural stone, a countertop material, and a decorative stone.

Gabbro

Type: Igneous

Gabbro is a dark-gray, coarse-grained intrusive igneous rock made predominantly of plagioclase and pyroxene. It is an important component of oceanic crust.

KEY FACTS

Gabbro is a plutonic rock that is dark gray to black.

- **grain size:** Coarse
- **texture:** Crystalline
- **composition:** Plagioclase + pyroxene ± olivine

Gabbro is a mafic rock, meaning it is silica-poor and iron- and magnesium-rich. It contains plagioclase feldspar plus pyroxene, usually augite, sometimes with olivine and/or hornblende. Gabbros with predominant orthopyroxene are called norites. Plagioclase in gabbro tends to be darker gray than in granitic rocks. Gabbro is coarse grained because it cools slowly at depth instead of erupting near or on the surface like basalt. It is an important component of oceanic crust. Exposures of gabbro can indicate rock formations that originated as ocean floor but were thrust onto continents by movement of the Earth's tectonic plates.

Anorthosite

Type: Igneous

Anorthosite is more than 90 percent plagioclase feldspar. It is uncommon, found in isolated exposures in some mountain belts and in ancient cores of the continents.

KEY FACTS

Anorthosite has interlocking grains of plagioclase feldspar, some with iridescent labradorite.

+ grain size: Coarse

+ texture: Crystalline; some with plagioclase grains in a dark-colored matrix

+ composition: 90 percent plagioclase, 10 percent ferromagnesian minerals

Anorthosite is an uncommon plutonic rock made almost entirely of plagioclase feldspar. Depending on the feldspar's color, anorthosites can be light gray or bluish. Anorthosite crystals are typically quite large. Some rock samples brought back from the moon are anorthosite. The origins of anorthosite are still debated among geologists. One hypothesis is that it formed from accumulation of plagioclase crystals that floated to the top of a magma chamber and then rose to shallower levels in the crust as a buoyant crystal mush. Anorthosite over 1.1 billion years old is exposed in the Adirondack Mountains of New York.

Peridotite

Type: Igneous

Peridotite is the dominant rock of the Earth's upper mantle. It is a dense, iron- and magnesium-rich intrusive igneous rock made of olivine and pyroxene.

KEY FACTS

Most peridotites have a reddish weathering surface.

+ grain size: Medium to coarse

+ texture: Crystalline; some layered; olivine often altered to serpentine

+ composition: Olivine + pyroxene ± hornblende + accessory minerals, including chromite

Peridotite is an ultramafic rock, meaning it is poor in silica and rich in iron and magnesium, resulting in a dense, often dark-colored rock. Many peridotite outcrops are slabs or pieces of mantle rock thrust onto the continents by large-scale movement of the Earth's plates during the construction of mountains or brought to the surface by volcanic eruptions. Peridotite comes in several subvarieties and many different textures, depending on the relative amounts of olivine and pyroxene, their chemical makeup, and the rock's magmatic history. Most peridotite outcrops contain secondary minerals including serpentine and talc.

Dunite

Type: Igneous

Dunite is a type of peridotite made of more than 90 percent olivine. Fresh exposures are a beautiful green, but most outcrops have a tan or brown weathering surface.

KEY FACTS

Dunite is an intrusive igneous rock of equigranular olivine with black grains of chromite as a common accessory.

- **grain size:** Medium to coarse
- **texture:** Crystalline, typically equigranular
- **composition:** More than 90 percent olivine

Dunite is an uncommon rock that is made almost entirely of olivine. Fresh, unaltered dunites have an equigranular texture of interlocking, glassy green olivine grains. These rocks may include black grains of chromite either dispersed with the olivine or aggregated in chromite layers. A significant supply of commercial chromium comes from chromite concentrations in dunite. Dunite is named after Dun Mountain in New Zealand, where it is part of a group of rocks that were formerly pieces of the ocean floor. Most dunite outcrops have a tan or brown weathering surface. In others, the olivine has been partially or completely altered to serpentine.

Rhyolite

Type: Igneous

Rhyolite is the extrusive equivalent of granite. It is a light-colored, fine-grained volcanic rock that forms from explosive eruptions of a highly viscous lava.

KEY FACTS

Rhyolite is a light to medium gray or pink volcanic rock with occasional flow banding; blocks are brittle with a flinty appearance.

+ grain size: Fine

+ texture: Can be glassy or porphyritic

+ composition: Same as granite

Rhyolite is an extrusive igneous rock that forms from the volcanic eruption of a felsic (silica-rich) magma. Rhyolitic lava is extremely viscous and sticky, so it does not flow very far or accumulate into typical volcanic cones; instead, it builds rounded lava domes. Rhyolitic eruptions are explosive and produce a lot of ash. Rhyolites' appearances include very fine-grained, glassy varieties, and porphyritic varieties with visible quartz. Rhyolite has few dark, iron- and magnesium-rich mineral phases, so it is typically light in color and weight. The countertop and tile stone known as *porfido trentino* is a rhyolite from Italy.

Pumice

Type: Igneous

Pumice is a volcanic rock characterized by many cavities of various sizes called vesicles. Pumice forms when the lava in a volcanic eruption turns frothy from the expulsion of water and gases.

KEY FACTS

A volcanic rock with many holes (vesicles), pumice floats in water. It is typically light in color.

+ grain size: None to very fine

+ texture: Rough surface, vesicular

+ composition: Varies; commonly rhyolite or andesite

Pumice is made of volcanic glass that is extremely porous due to its many gas bubbles, called vesicles. Pumice forms from magma that is saturated in gases and water. During volcanic eruptions, these are released, creating a frothy foam-like lava. This lava cools so quickly that no crystals are able to grow, and the resulting rock is called a glass. Pumice is usually light in color and typically has the same composition as a rhyolite or andesite, depending on the composition of the magma from which it formed. Pumice is used as an abrasive material, including a bath accessory for smoothing rough skin.

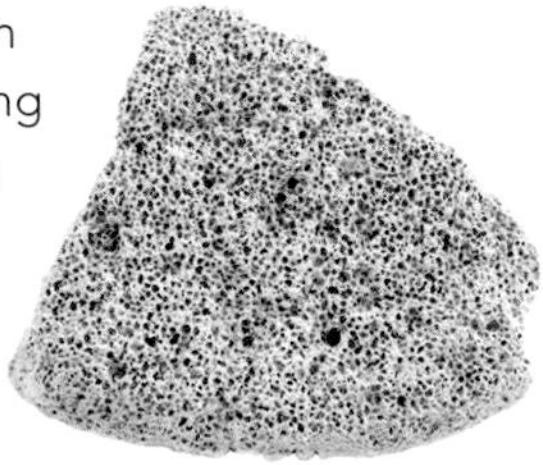

Obsidian

Type: Igneous

Obsidian is volcanic glass that forms from viscous lava, typically in rhyolite lava domes. The brittle nature of obsidian and its conchoidal fracture make it an ideal material for sharp tools.

KEY FACTS

This shiny volcanic glass has a conchoidal fracture; banding and inclusions are common; colors are variable.

+ grain size: None to very fine

+ texture: Glassy

+ composition: Can vary; commonly rhyolite

Obsidian is glassy and brittle with a conchoidal fracture, meaning it breaks irregularly along curved surfaces. Paleolithic hunter-gatherers exploited these properties, chipping conchoidal flakes from obsidian pieces to create sharp projectile points and other tools. Obsidian forms in lava domes from lava so sticky and viscous that it is cooled too rapidly for minerals to crystallize, resulting in a glass. Silica-rich lavas like rhyolite or dacite are the most viscous and, thus, most common form of obsidian. Obsidian is frequently opaque black, but it can have various colors and patterns, including intricately banded forms.

Tuff

Type: Igneous

Volcanic eruptions produce massive amounts of ash and other material known as tephra. If enough heat is retained after it falls to the ground, tephra will fuse together, forming tuff.

KEY FACTS

Tuff is a volcanic rock made of consolidated ash and other volcanic fragments; can be porous.

+ grain size: Variable

+ texture: Visible volcanic clasts; possible layering or bedding

+ composition: Can vary; commonly rhyolite or andesite

Material ejected into the atmosphere during a volcanic eruption is classified by size. Particles smaller than 0.08 inch (2 mm) are known as ash, particles between 0.08 and 2.5 inches (2 and 64 mm) are called lapilli, and anything larger is a volcanic bomb. Collectively, this material is called "tephra," the Greek word for ash. When tephra falls to the Earth and solidifies, it forms a rock called tuff. Tuffs can also form when tephra becomes slowly compacted and cemented into a rock. Tuffs vary in their composition and appearance. Most have visible volcanic fragments of different sizes and shapes, including many that are angular.

Dacite

Type: Igneous

Dacite is a volcanic rock of intermediate composition between rhyolite and andesite. It is the volcanic equivalent of granodiorite. Dacite is associated with highly explosive volcanic eruptions.

KEY FACTS

Dacite often has visible quartz and/or plagioclase; it can be light gray or dark gray to black.

+ grain size: Fine to medium

+ texture: Commonly porphyritic with phenocrysts of plagioclase, quartz, biotite, or hornblende

+ composition: Same as granodiorite

Dacite is a volcanic rock that is slightly less rich in silica than rhyolite, but like rhyolite, comes from a highly viscous lava and produces violently explosive volcanic eruptions. Dacite forms thick rounded lava domes on the surface of volcanoes, including the lava dome at Mount St. Helens. Dacite is named after a locality in the mountains of Romania, part of a region known during the Roman Empire as Dacia. Dacite is typically light gray, but some can be dark gray or even black. Dacite is common in volcanoes that form along continental subduction zones—margins of continents beneath which oceanic slabs sink.

Andesite

Type: Igneous

Andesite is a volcanic rock similar to basalt but with more silica. It is the volcanic equivalent of diorite. Andesitic volcanoes are steep-sided cones, common around the Pacific Rim.

KEY FACTS

A common igneous rock, andesite is typically medium to dark gray; can have greenish or reddish hues.

+ grain size: Fine to medium

+ texture: Commonly porphyritic with phenocrysts of plagioclase and/or pyroxene or hornblende

+ composition: Same as diorite

Andesite is an extrusive igneous rock that forms from the lava flows of stratovolcanoes. Stratovolcanoes are a type of composite volcano built from alternating layers of lava, ash, and cinders, resulting in a steep-sided, symmetrical cone. Some of the world's most beautiful mountains are andesitic stratovolcanoes, including Mount Fuji in Japan and Mount Shasta, Mount Hood, and Mount Adams in the Pacific Northwest. Andesite is named for the Andes Mountains of South America where the rock is also common. Andesite is typically porphyritic, with visible crystals of plagioclase or pyroxene set in a fine-grained groundmass.

Diabase

Type: Igneous

Diabase, also called dolerite, has the same composition as basalt and gabbro. These are distinguished by their grain size: basalt fine, diabase medium, and gabbro coarse.

KEY FACTS

Diabase forms dikes and sills; most outcrops have a light-brown weathered surface; fresh surfaces are dark gray or green overall.

+ **grain size:** Medium
+ **texture:** Crystalline; commonly porphyritic
+ **composition:** Plagioclase + pyroxene + olivine

Diabase, also known as dolerite, is a dark-colored rock that forms dikes, sills, and other relatively small igneous bodies. Diabase is compositionally identical to basalt and gabbro, and all three rock types come from the same type of magma. Basalt, however, forms when the magma erupts onto the Earth's surface, and gabbro forms when the magma cools slowly at great depths. Some outcrops called traprock are made of diabase. Diabase traprock near Washington, D.C., is quarried and crushed for use in concrete and as road base material. The Palisades of the Hudson River is an enormous sill made of diabase.

Basalt

Type: Igneous

Basalt forms extensive lava flows worldwide. The oceanic crust is composed primarily of basalt, and the lunar "maria"—the dark regions visible on the moon—are also basalt.

KEY FACTS

A common dark-colored volcanic rock, basalt is typically black, gray, or brown; surfaces vary from smooth to sharp and cindery.

+ **grain size:** Fine

+ **texture:** Can be porphyritic

+ **composition:** Plagioclase + pyroxene + olivine

Basalt is one of the most common rock types. It is a mafic igneous rock with an aphanitic texture, meaning its grains are so fine that they cannot be distinguished with the naked eye; thus, the rock looks relatively uniform. Porphyritic basalts are exceptions; they contain visible plagioclase or olivine crystals surrounded by a fine-grained groundmass. Basalt lava flows cover much of the Earth's surface, including the ocean floor, volcanoes in Hawaii and Iceland, and large regions of continents known as flood basalts. Columbia River Basalts are flood basalts that extend across Washington, Oregon, and Idaho.

Sandstone

Type: Sedimentary

Sandstone is made of compacted and cemented sand-size particles. Sandstones vary depending on composition of the grains, type and amount of cement, and depositional environment.

KEY FACTS

Sandstone is a sedimentary rock made of sand-sized grains.

+ grain size: Medium

+ texture: Clastic

+ composition: Variable; common components include quartz, feldspar, mica, and rock fragments.

Sandstone is one of the most common sedimentary rocks. It is made of particles or clasts that fall within the size range that classifies sand, 0.0008 to 0.08 inch (0.02 to 2 mm). This is a broad definition—the only parameter being grain size—thus, the rocks that are identified as sandstones can vary widely. Depending on how rounded or angular their grains are and the amount of compaction of the sediment, sandstones can be quite porous and thus serve as important reservoirs worldwide for resources such as water and hydrocarbons. Sandstones are used for building facades and unpolished floor tiles. Decorative flagstone is typically sandstone.

Quartz Arenite

Type: Sedimentary

Quartz arenite is a sandstone composed almost entirely of quartz. Quartz arenites form from accumulations of quartz sand, commonly in windblown deserts, beaches, and high-energy rivers.

KEY FACTS

This sedimentary rock is made of sand-size grains of quartz; cement is often quartz but can also be calcite or hematite.

+ **grain size:** Medium
+ **texture:** Clastic
+ **composition:** More than 90 percent quartz

An arenite is a texturally mature sandstone, meaning it is of consistent grain size and shape, and contains few to no clay particles between the grains. Its grains are predominantly quartz. Sediments mature during weathering, transportation, and deposition, as less resistant minerals break down and more resistant minerals become rounded. Quartz arenites form from sediment deposited far from the source, typically in environments such as sand dunes, coastlines, and rivers. These sandstones can take on different colors, commonly cream, orange, or red, depending on small amounts of iron oxide coating the sand grains.

Graywacke

Type: Sedimentary

A wacke is a variety of sandstone with mud or clay in the matrix that surrounds the sand grains. A graywacke is a type in which many of the sand grains are rock fragments.

KEY FACTS

An immature, poorly sorted sandstone, graywacke is typically dark.

+ grain size: At least 50 percent sand size with fine-grained matrix

+ texture: Clastic, poorly sorted

+ composition: Rock fragments, feldspar, quartz, minerals rich in iron and magnesium, clay, and mud

Unlike arenites, wackes are sandstones that are not well sorted—they contain a mixture of grain sizes, from clay to sand, as well as grain shapes from angular to rounded. The term "graywacke" refers to wackes with a composition that includes numerous rock fragments as well as minerals such as micas. The minerals and fragments are fragile, so their preservation in a sandstone indicates that the sediment source cannot be far. Geologists interpret many graywackes as deposits of turbidity currents—strong currents in water moving down a slope. These rocks are sometimes called turbidites instead of graywackes.

Arkose

Type: Sedimentary

An arkose is a type of sandstone made up of at least 25 percent feldspar. Arkoses typically also contain quartz and mica grains and rock fragments.

KEY FACTS

A sedimentary rock, arkose is made of sand-size grains of feldspar and quartz; cement is commonly quartz or calcite.

- **grain size: Medium**
- **texture: Clastic**
- **composition: More than 25 percent feldspar**

An arkose or arkosic sandstone contains a moderate to high percentage of feldspar grains. Feldspar is susceptible to breakdown via chemical weathering, so its presence in a sandstone indicates that the sediment source is not far from its depositional environment. Arkoses thus tend to form relatively close to mountains or other uplands from which feldspar-rich sediment, typically from granitic rocks, is shed. In fact, geologists sometimes use arkosic sandstones in the sedimentary record as an indication of previous mountain belts. Arkoses are often pink or red, but can also have gray or even greenish hues.

Limestone

Type: Sedimentary

Limestone is made up mainly of calcite (calcium carbonate). It is primarily a marine sedimentary rock, though some freshwater limestones are known to occur.

KEY FACTS

A light-colored sedimentary rock, limestone is often fine grained; some formations are rich in marine fossils. It fizzes in contact with hydrochloric acid.

+ grain size: Very fine to coarse

+ texture: Variable

+ composition: Primarily calcite

Calcium carbonate can chemically precipitate out of solution in sea or lake water, settling to the bottom, accumulating as a sediment, and forming limestone. Limestone may also form from accumulated animal shells made of calcium carbonate. Thus, fossils are common in limestones; some limestone formations are made entirely of fossil shells. Some limestones contain small, spherical beads of calcite called oolites that develop when calcium carbonate precipitates around a particle such as a sand grain. Limestone forms cliffs in arid environments, but forms caves and other karst landforms in humid environments.

Dolomite/Dolostone

Type: Sedimentary

Dolomite is similar to limestone but with a high percentage of the mineral dolomite—calcium magnesium carbonate—instead of calcite. Dolomite rock is sometimes called dolostone.

KEY FACTS

Although it is similar to limestone, dolomite does not fizz as actively in hydrochloric acid; dolomite's weathering surface is a buff color, whereas limestone tends to be grayer.

+ **grain size:** Fine to medium
+ **texture:** Compact, relatively homogeneous
+ **composition:** Primarily dolomite

Dolomite, also known as dolostone, is similar to limestone and is often found in association with limestone. Most dolomite originated as limestone or lime mud, and became dolomite when calcite (calcium carbonate) was replaced by dolomite (calcium magnesium carbonate), a process called dolomitization. Dolomite and limestone can be difficult to distinguish in the field. Dolomite tends to weather to yellowish beige or brown, whereas limestone tends to be gray. Dolomite is named after French geologist Déodat de Dolomieu, who first described it in the 1700s from cliffs in the Italian Alps now called the Dolomites.

Travertine

Type: Sedimentary

Travertine is a type of limestone formed via direct chemical precipitation of calcite out of solution. This commonly occurs in association with the waters around hot springs.

KEY FACTS

Deposits of travertine are associated with caves, hot springs, and geysers; typically layered with light colors; fossils are common.

+ grain size: Very fine

+ texture: Massive, nonclastic, porous

+ composition: Calcite (calcium carbonate)

Travertine is a beautiful ivory- to peach-colored stone that forms around hot springs when calcium carbonate falls out of solution. This occurs when carbon dioxide is released as heated waters rise from depth or are agitated as in waterfalls and cascades. The results are porous, finely layered deposits of calcium carbonate. Travertine is a popular building stone. Much of Rome, which is surrounded by volcanic thermal springs, was constructed with travertine, including the Colosseum. Travertine is also used to clad modern buildings including the Getty Center in Los Angeles and Lincoln Center in New York.

Chalk

Type: Sedimentary

Chalk is a variety of limestone made of tiny skeletons of minute marine organisms. The skeletons, composed of calcium carbonate, accumulate on the floors of shallow seas.

KEY FACTS

White, pure limestone chalk is relatively soft, made of microscopic shells with a carbonate mud matrix.

+ grain size: Fine

+ texture: Bioclastic

+ composition: Calcite (calcium carbonate)

Chalk is a relatively pure limestone made of accumulated shells of marine microorganisms cemented together with a carbonate mud. Many chalk deposits formed during the Cretaceous period, from about 142 to 65 million years ago, a time when global sea levels were high and parts of the continents were flooded with shallow seas. "Cretaceous" comes from the Latin word *creta*, meaning chalk. Many small, calcareous marine organisms thrived in these environments, and deposits of their skeletons built up on the sea floor to form layers of chalk. Chalk formations commonly contain nodules of chert known as flint.

Conglomerate

Type: Sedimentary

Conglomerates are sedimentary rocks composed of particles larger than sand grains, including pebbles, cobbles, and boulders. They form in various depositional environments.

KEY FACTS

Coarse-grained clasts of conglomerates are surrounded by a fine-grained matrix; relative proportions of clasts and matrix vary; size, shape, and composition of clasts also vary.

+ grain size: Larger than 0.08 in (2 mm)

+ texture: Clastic

+ composition: Variable

Sedimentary rocks made of particles larger than sand grains are known as conglomerates. The particles vary from rounded river pebbles to angular fragments of a debris flow. Conglomerates represent high-energy deposition including river- or beach-deposited gravel beds, alluvial fans, glacial till deposits, and mudslides. Conglomerates are classified by the relative proportion of clasts and matrix—the fine-grained material cementing the rock together. Conglomerates in which the clasts touch are known as clast supported. Those in which the clasts are surrounded by matrix and not touching are called matrix supported.

Breccia

Type: Sedimentary

Breccias contain coarse, angular particles surrounded by finer grains. The term "breccia" is used to describe sedimentary formations and rocks with fault-related or igneous origins.

KEY FACTS

Breccias show angular, broken fragments surrounded by a fine-grained matrix.

+ grain size: Larger than 0.08 in (2 mm)

+ texture: Clastic

+ composition: Variable

Breccias are similar to conglomerates but are distinguished by their angular, broken rock fragments instead of rounded clasts. The angular clasts can come from a single source or from multiple rock types. Like other sedimentary rocks, the cement is commonly calcareous or siliceous. Breccias form in various environments. Sedimentary breccias can form from angular fragments of an underwater debris flow, deposits of a landslide, or other mass-wasting events. Fault breccias form when rocks are fragmented during slip along a brittle fault zone. Volcanic breccias form from angular fragments ejected during an eruption.

Shale

Type: Sedimentary

Shale is a sedimentary rock made of silt- and clay-size grains—particles smaller than sand. It is distinguished by its fissility, meaning its tendency to split into thin layers parallel to bedding.

KEY FACTS

A clastic sedimentary rock, shale splits easily into layers.

+ grain size: Very fine

+ texture: Fine grains, fissile

+ composition: Variable; common mineral grains include quartz, feldspars, mica, calcite, and clays.

Sedimentary rocks made of grains finer than sand—a mixture of silt and clay—are generally known as mudrocks. Shale is a mudrock characterized by its property of easily breaking in one direction parallel to bedding. This is called fissility, and is formed by alignment of platy minerals such as micas and clays. Shale forms from compaction of fine-grained sediment, a process that typically occurs in still or slow-moving water as in bogs, deltas, and deep regions of the continental shelf. Fossil preservation is enhanced in these sediments and quiet depositional environments; thus, shales are common sources of fossils.

Mudstone

Type: Sedimentary

Mudstone is a fine-grained sedimentary rock made up of a mixture of clay- and silt-size particles. Unlike shale, mudstone is massive and does not easily split into layers.

KEY FACTS

A fine-grained sedimentary rock, mudstone does not easily split into layers.

+ grain size: Very fine

+ texture: Massive, not laminated

+ composition: Variable; common mineral grains include quartz, feldspars, mica, calcite, and clays.

Mudstone is a common sedimentary rock. It is clastic, but its particles are generally too small to see with the naked eye. Both mudstone and shale are made of compacted mud; the difference lies in the way these rocks break. Mudstone has a tendency to break into irregular blocks, whereas shale consistently breaks along a plane that is parallel to bedding. Shales and mudstones come in a variety of colors, most commonly gray, but also red, green, and purple. The colors usually indicate the presence of iron—red when iron is oxidized, green when iron is partially reduced.

Gneiss

Type: Metamorphic

Gneiss (pronounced nice) is a metamorphic rock characterized by compositional bands of separated light and dark minerals. These bands define a coarse layering or foliation in the rock.

KEY FACTS

This foliated metamorphic rock is defined by compositional banding, with few platy or needlelike minerals.

+ grain size: Medium to coarse

+ texture: Crystalline; compositional banding

+ composition: Variable; quartz, feldspar, and hornblende are common components.

Gneisses form from metamorphism of preexisting rocks that are typically rich in quartz and/or feldspar, such as granitic rocks and sandy sedimentary formations. Gneisses with an igneous parent rock are classified as orthogneisses; those with a sedimentary parent are called paragneisses. Gneisses form when the parent rocks are exposed to high temperature and pressure in the Earth's crust, usually by burial during mountain building. Gneisses' compositional bands are layered segregations of light-colored minerals such as quartz and feldspar from dark-colored minerals such as hornblende.

Migmatite

Type: Metamorphic

Migmatites, or migmatitic gneisses, are rocks that have partially melted at high temperatures and pressures, resulting in a hybrid rock with both igneous and metamorphic properties.

KEY FACTS

These rocks are heterogeneous, with segregations of light- and dark-colored minerals usually in irregularly folded layers.

+ **grain size:** Medium to coarse

+ **texture:** Crystalline

+ **composition:** Variable; light layers commonly granitic; dark layers generally contain hornblende and biotite.

Migmatites are named from the Greek *migma,* meaning mixture. They are mixtures of igneous and metamorphic components, formed when metamorphism progresses past the point at which the rock begins to melt. The melting is incomplete, producing segregations of melt that become swirly, light-colored igneous layers surrounded by the dark-colored residual metamorphic rock. Migmatites form under very high temperatures and pressures deep in the Earth's crust, as in the cores of large mountain ranges. These rocks are used as decorative stones, and they are often incorrectly called granites.

Amphibolite

Type: Metamorphic

Amphibolite is a rock rich in amphibole—typically hornblende or actinolite—often with plagioclase. Some amphibolites are massive; others have fine foliation or bands.

KEY FACTS

Colored black or dark green, amphibolite is often associated with gneisses and/or schists.

+ grain size: Medium to coarse

+ texture: Variable; can be massive, foliated, or banded

+ composition: Hornblende or actinolite plus plagioclase, sometimes garnet, and other minerals

Amphibolites are widespread metamorphic rocks made predominantly of amphibole plus plagioclase. Some amphibolites display an easily recognizable salt-and-pepper appearance with grains of black hornblende and white plagioclase. Amphibole grains are typically aligned in a weak to moderate foliation. The protolith, or parent rock, of amphibolites is generally difficult to determine. Most amphibolites form by metamorphism of mafic igneous rocks such as basalt and gabbro, though some originate as sedimentary rocks. Amphibolite is used as a building and decorative stone, which is often called "black granite."

Mylonite

Type: Metamorphic

Mylonites are hard, compact, foliated rocks with very fine grains. They commonly display a strong linear fabric defined by elongate minerals.

KEY FACTS

Strong deformational fabrics, compact texture, and fine grain size distinguish mylonites.

+ grain size: Very fine; some coarse relict grains common

+ texture: Well-developed foliation and lineation

+ composition: Variable

Mylonites form when rocks are deformed by movements along a fault. They typically form in deep zones of faults where rocks deform in a ductile fashion, like warm plastic, rather than breaking brittlely. As a rock is sheared, its minerals are pulverized on a microscopic scale into smaller grains. Harder minerals like quartz and feldspars can resist complete pulverization, resulting in larger, rounded, or oval grains surrounded by the fine-grained material. If there is significant shearing, even the resistant grains will be reduced to the fine matrix. Most mylonites form along discrete shear zones.

Schist

Type: Metamorphic

Schist is a common metamorphic rock characterized by its schistosity, which is a fine layering formed by the preferred orientation of platy and needlelike minerals, typically micas.

KEY FACTS

A finely layered (schistose) rock, schist has more than 50 percent sheetlike minerals, commonly muscovite, biotite, and other micas, and/or chlorite.

+ grain size: Medium to coarse

+ texture: Well-developed schistosity

+ composition: Mica- or chlorite-rich, otherwise variable

The term "schist" comes from the Greek *schistos,* meaning "divided." Schists typically split along parallel planes defined by the alignment of platy minerals such as micas. These planes are a type of foliation called schistosity. Schists form from metamorphism of fine-grained sedimentary rocks such as mudstones and shales. These rocks typically first form slate, then phyllite, an intermediate rock; eventually, as temperatures and pressures increase, they form schist. Schists can also develop from metamorphism of other fine-grained rocks such as volcanic tuffs. Some schists display tight chevron-type folds.

Garnet Schist

Type: Metamorphic

Schists can be classified according to their significant mineral components; for example, a biotite schist or a muscovite-chlorite schist. Garnet schist or garnet-mica schist features garnet.

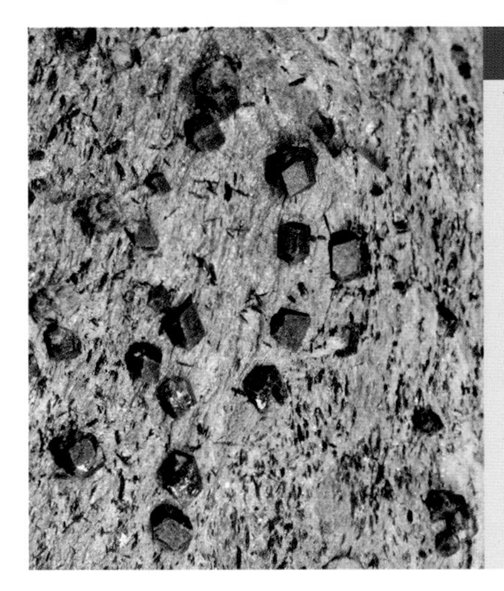

KEY FACTS

This form of schist has rounded garnet crystals, typically deep red, embedded in a scaly matrix of smaller, aligned micas.

+ **grain size:** Medium
+ **texture:** Well-developed schistosity; commonly larger than other minerals
+ **composition** Typically muscovite and garnet; can include biotite, staurolite, kyanite, or sillimanite

If a mudstone or shale is buried at depth by large-scale movements in the Earth's crust (for example, during the building of a mountain belt), the rock undergoes progressively higher pressures and temperatures. These conditions cause the rock to metamorphose into a schist. Types of minerals that grow in the schist depend on the peak pressure and temperature conditions reached by the rock. Thus, the mineral makeup of some schists is an important tool for scientists to understand the geologic history of a region. Garnet schists can indicate that a rock reached relatively high pressures and/or temperatures.

Greenschist

Type: Metamorphic

These fine-grained metamorphic rocks are rich in chlorite, epidote, or actinolite. The minerals give the rock an overall greenish hue.

KEY FACTS

Greenschist is a chlorite schist or other foliated metamorphic rock with chlorite and/or actinolite.

+ grain size: Medium

+ texture: Foliated

+ composition: Chlorite, epidote, actinolite

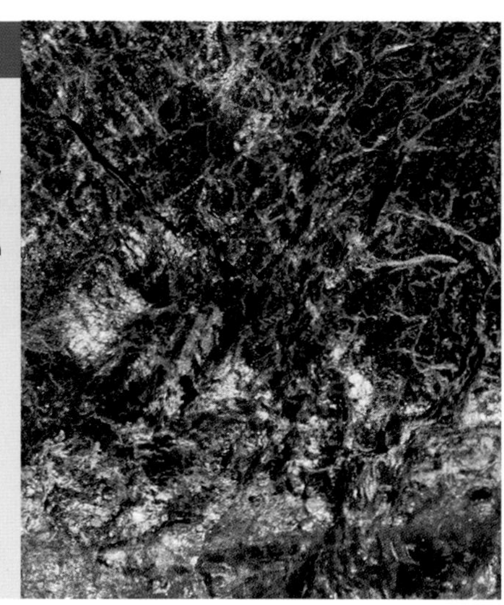

The term "greenschist" usually describes a chlorite schist, though it can refer to other foliated metamorphic rocks that are rich in green-colored minerals such as chlorite, actinolite, and epidote. Greenschists have a foliation defined by the alignment of chlorite grains. Most greenschists form when a fine-grained mafic volcanic rock such as a basalt is metamorphosed at low temperatures and pressures. The term "greenschist" also describes metamorphic conditions. Greenschist metamorphism occurs during low to medium pressures and temperatures. For example, a slate is not a greenschist, but it develops under greenschist conditions.

Blueschist

Type: Metamorphic

Blueschist is a metamorphic rock that typically has a blue hue from the presence of the mineral glaucophane, a sodium-rich amphibole. The term also describes a set of metamorphic conditions.

KEY FACTS

Blueschist is a foliated metamorphic rock with a bluish color.

+ **grain size:** Medium

+ **texture:** Foliated, not necessarily schistose

+ **composition:** Variable; blue color comes from glaucophane

A blueschist rock forms under blueschist metamorphic conditions, but not all rocks that experience blueschist metamorphism turn into blueschists. Confused yet? The complication comes from two different uses for the word "blueschist," a similar scenario to the use of the word "greenschist." The rock type called blueschist is a metamorphic rock that contains the bluish amphibole glaucophane. The term "blueschist" can also refer to a set of metamorphic conditions characterized by high pressures and low temperatures. These conditions occur along subduction zones, where one of the Earth's plates is pulled down beneath another.

Slate

Type: Metamorphic

Slate is a fine-grained metamorphic rock that splits easily along a well-developed foliation. Slate is commonly used for roofing tiles as well as for other decorative stones.

KEY FACTS

Typically gray or greenish, slate is a brittle rock with well-developed foliation.

+ grain size: Fine

+ texture: Compact, well foliated

+ composition: Rich in micas and quartz, though often too fine-grained to distinguish individual minerals

Slate is a common metamorphic rock characterized by its ability to split into very thin sheets or plates, some of which are quite large. Slate is actively quarried for use as floor tiles, hearths, and roofing tiles. The original blackboards in classrooms were made of single sheets of slate, though most blackboards today are composed of man-made materials. Slate forms via the metamorphism of mudstones, shales, or tuffs at low pressures and temperatures. If metamorphism progresses, mica crystals grow larger, turning a slate into a schist. Quartz veins and small faults are common in slate outcrops.

Marble

Type: Metamorphic

The metamorphism of limestone or dolomite results in formation of marble, a rock that is composed almost entirely of recrystallized, interlocking grains of calcite or dolomite.

KEY FACTS

A dense stone, marble is usually pure white, gray, or yellow; other varieties are pink, blue, or black.

+ grain size: Fine to coarse

+ texture: Crystalline; often massive; some show banding

+ composition: Calcite and/or dolomite; accessory minerals can include micas, graphite, and serpentine.

Some of the most famous sculptures and buildings are created out of marble, a type of metamorphic rock. Michelangelo's "David" is carved out of an Italian stone known as Carrara marble. The Taj Mahal in India is clad almost entirely in a marble known as White Makrana. In the United States, the Supreme Court Building and the Lincoln Memorial are built with slabs of Alabama marble. Marble is white when it is pure calcite. More commonly, marbles contain impurities and other minerals resulting in different colors and textures. Some coarse-grained marbles can look similar to quartzite, but marble is much softer.

Serpentinite

Type: Metamorphic

Serpentinite is composed of serpentine, often with chromite or magnetite and veins of calcite, dolomite, or talc. It forms by alteration and metamorphism of ultramafic rocks.

KEY FACTS

Light yellow-green, dark green, or black, serpentinite is typically brittle with a greasy or waxy feel.

+ grain size: Fine

+ texture: Variable; brecciated forms, veins common

+ composition: Primarily serpentine; other minerals may include chromite, magnetite, and talc.

Serpentinite is a striking rock, commonly streaked with various colors from yellow-green to black, with a glossy surface that is greasy or even soapy to the touch. Serpentinite forms when olivine-rich rocks like peridotite are altered in the presence of water, a process known as serpentinization. The formation of serpentinites occurs at low metamorphic grades. Serpentinite is a popular stone used for floor tiles and countertops in buildings, and as an ornamental stone. The popular decorative stone known as *verd antique* ("antique green") is a serpentinite with a brecciated or angular, broken texture.

Quartzite

Type: Metamorphic

Quartzite is extremely hard and compact, formed by metamorphism of quartz-rich sandstones or chert at elevated temperatures and pressures.

KEY FACTS

Hard, compact, and quartz-rich, quartzite has a distinctive sugary appearance. It is commonly white or gray.

+ **grain size:** **Medium**

+ **texture:** **Compact, interlocking grains**

+ **composition:** **Primarily quartz; other minerals may include micas and/or hematite.**

Quartzite forms when quartz grains of the protolith—the parent rock before metamorphism—are heated and squeezed together until they form new grains. This process creates a homogeneous mosaic of interlocking grains, resulting in a dense, very hard stone. Textures from the protolith—rounded quartz grains surrounded by fine quartz cement, for example—are destroyed. Quartzites are distinguished by their hardness, conchoidal fracture, and sugary texture in outcrop. The decorative stone aventurine is quartzite. Green aventurine, used in jewelry, is colored by a chromium-rich mica known as fuchsite.

Plants

Plants are among the most diverse groups of organisms. Their fossils help scientists understand the evolution of life as well as changes in the Earth's climate through geologic time.

KEY FACTS

Preserved parts of plants include leaves, bark, seeds, spores, and pollen.

+ time period: About 450 million years ago to present

+ size: Variable

+ where to find: Fine-grained sedimentary rocks

Plant fossils are the geologic record of ancient plants. The first land plants appeared about 450 million years ago. Complete fossils are rare; most are fragments such as leaves, cones, and stems. Most plants decompose rapidly; to become fossilized, specific conditions have to occur. The plant must be buried by sediment quickly enough to shield it from decomposition, but gently enough to preserve its integrity. Thus, most plant fossils are found in fine-grained sedimentary rocks such as shales and mudstones. Fossil plants, especially leaves, spores, and pollen grains, are important to our understanding of past climates.

Petrified Wood

Petrified wood is a plant fossil preserved by the infiltration of minerals around the organic material, a process called permineralization. Most petrified wood is made of quartz or calcite.

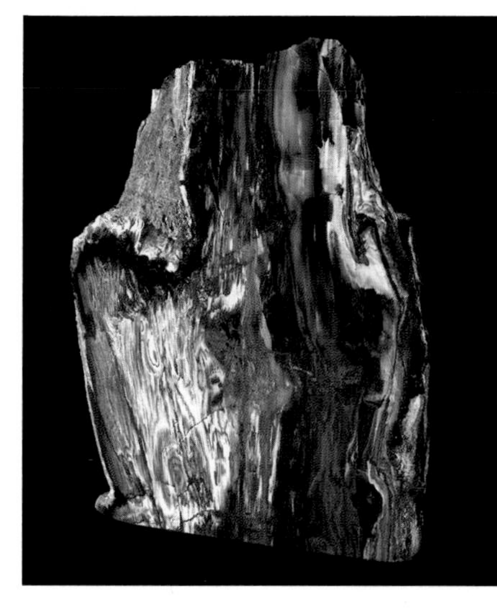

KEY FACTS

Three-dimensional replicas of woody plants, including trunks, branches, and bark; harder and more brittle than unfossilized wood.

+ time period: About 400 million years ago to present

+ size: Variable

+ where to find: Eroded out of sedimentary rock formations

Ancient plants fossilized by permineralization are called petrified wood. The pore spaces around and within the plant's organic material are filled by minerals, usually microcrystalline quartz. The organisms retain their original structures, so permineralized fossils are three-dimensional forms, not casts or impressions. Petrified wood is found throughout the U.S. and Canada, especially in western states. In Arizona's Petrified Forest National Park, entire trees over 200 million years old were fossilized in buried logjams of ancient rivers. Petrified wood can be colorful due to the presence of iron, carbon, and manganese.

Graptolites

Graptolites are a group of small, free-floating marine organisms that flourished during the early and middle Paleozoic eras. Graptolite fossils are found worldwide, mostly in black marine shales.

KEY FACTS

These small, twiglike fossils have saw-blade shapes, some in spirals.

+ time period: 500–315 million years ago

+ size: Typically 1–4 in (2.5–10 cm); some as long as 8–10 in (20–25 cm)

+ where to find: Marine shales of Ordovician, Silurian, and Devonian periods

Graptolites are a group of relatively simple marine animals that lived during the early explosion of life in the Earth's oceans beginning about 500 million years ago. Graptolites are wormlike animals that lived in colonies floating through seawater like plankton. Graptolites constructed tubelike shells out of soft collagen, and the tubes were linked to form branching colonies. Many graptolite fossils resemble plant leaves or branches of plants. Others take the shape of tiny saws, tuning forks, and spirals. Graptolites are typically found in black marine shales associated with calm waters of the outer continental shelf.

Bryozoans

Bryozoans are colonial aquatic organisms that resemble small corals. They are made of many small individuals living in compartments of a calcareous skeleton.

KEY FACTS

Lacy or fanlike fossils resemble moss or corals.

+ **time period:** 488 million years ago to present

+ **size:** Individuals are less than 1 mm; colonies range from millimeters to meters.

+ **where to find:** Shallow-marine limestone formations; also shales and mudstones

Bryozoans are colonial organisms that secrete calcareous shells with many compartments. Each compartment contains a body cavity with a retractable food-gathering arm called a lophophore. The lophophore is a group of tentacles armed with beating fibers called cilia that gather food particles from the seawater. Most bryozoans attach themselves to the sea bottom onto a rocky substrate or the discarded shell of another organism. They take many different forms, including sheetlike crusts, netted fans, and small branching trees. They build reefs. Bryozoan fossils are most common in shallow marine limestone formations.

Corals

Corals live in tropical marine settings and are the principal component of reefs. Corals are ancient animals, with fossil specimens dating back more than 500 million years.

KEY FACTS

Skeletons of coral have various shapes including horns, brains, spirals, branches, and disks.

+ time period: 540 million years ago to present

+ size: Variable; millimeters to meters

+ where to find: Limestones

Corals belong to the phylum Cnidaria (organisms with stinging cells), which includes sea anemones and jellyfish—simple, soft-bodied marine organisms. Corals construct a skeleton made of calcium carbonate and are well preserved as fossils. Living corals consist of a soft body cavity called a polyp with a single opening, surrounded by tentacles that gather food from the water. The polyps are attached to a hard skeleton that they construct throughout their life. Coral fossils display many shapes, including disks, cylinders, spirals, and brain-like domes. Because of their limited ecological tolerances, corals are indicators of ancient environmental conditions.

Trilobites

Trilobites are ancient animals, now extinct, that dominated the world's oceans during the Cambrian period. Related to crustaceans, they had a hard outer shell that is well preserved.

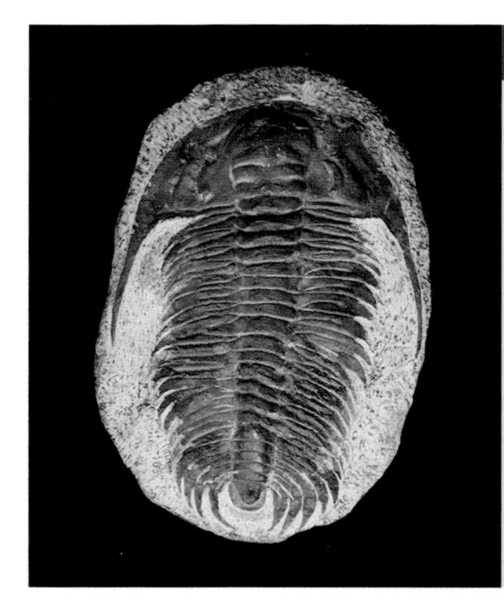

KEY FACTS

Body is segmented, and has an armored shell and crescent-shaped head.

+ time period: 540–20 million years ago

+ size: Typically 1–4 in (2.5–10 cm)

+ where to find: Fine-grained marine sedimentary rocks; famous locales are Burgess Shale in British Columbia, Beecher's Trilobite Bed in New York, and Wheeler Shale in Utah.

Trilobites are among the earliest arthropods—a classification of invertebrate animals with exoskeletons, including insects, spiders, and crustaceans. Most trilobites were 1 to 4 inches (2.5 to 10 cm) long, though one of the largest trilobites, found on the shore of Hudson Bay, measures 28 inches (71 cm). Trilobites were prolific and geographically dispersed, roaming worldwide Paleozoic oceans in deep and shallow water environments. These factors, plus the easy preservation of their shells, resulted in trilobite fossils being both common and widespread around the globe. Trilobite fossils are important time markers for sedimentary formations.

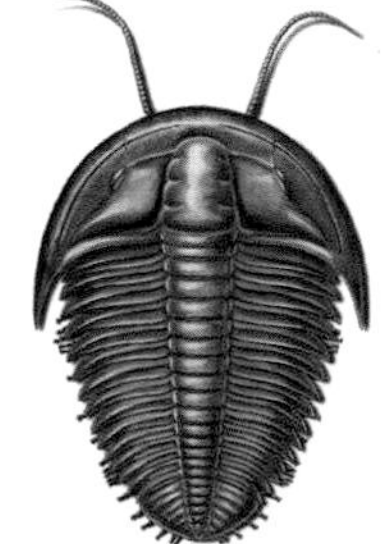

Brachiopods

Brachiopods, also known as lampshells, are marine animals with two shells connected at a hinge. At first glance, they appear similar to bivalves, though these animals are unrelated.

KEY FACTS

The two shells of brachiopods are symmetrical along the midline, perpendicular to the hinge, with the bottom shell typically larger than the top shell.

+ time period: 540 million years ago to present

+ size: Most are about 0.25–4 in (0.6–10 cm); largest up to 8–12 in (20–30 cm)

+ where to find: Limestones and marine shales

Brachiopods are marine organisms with two shells, similar to clams and mussels. Clams and mussels, however, are bivalves, unrelated to brachiopods and different in many ways. In brachiopod shells, the left half is a mirror image of the right; they are commonly asymmetrical top to bottom, with the bottom larger than the top. Though only a few hundred species live today, brachiopods dominated the world's oceans approximately 540 to 252 million years ago, when tens of thousands of species flourished. Many species died off during a mass extinction event approximately 250 million years ago at the end of the Permian period.

Bivalves

Bivalves are marine and freshwater organisms with two shells joined along a hinge. Fossil bivalves are found worldwide from all time periods since the Cambrian.

KEY FACTS

The two shells of bivalves are symmetrical parallel to their hinge; the valves are mirror images of each other.

+ time period: 540 million years ago to present

+ size: Most are 0.4–4 in (1–10 cm); some are as large as 3.3 ft (1 m) or more.

+ where to find: Limestones and shales

Bivalves are aquatic animals with two shells that meet along a hinge. These include living clams, oysters, mussels, and scallops. Bivalves have a muscular foot that emerges from the shell and allows some species to burrow into the sand or mud. Other bivalves anchor themselves to rocks or coral reefs. Others, such as scallops, swim by clapping their shells together, forcing out water to propel them. Fossil bivalves are common, and they come in a variety of shapes and sizes. The symmetry of their two shells is key to their identification. In contrast, brachiopod symmetry is perpendicular to the orientation of the hinge.

Gastropods

Gastropods are snails, most with coiled, hard shells made of calcium carbonate. They are found in a range of habitats, including marine and freshwater environments as well as terrestrial habitats.

KEY FACTS

Snails and sea slugs are gastropods. Snail shells come in various shapes with single chambers. Sea slugs have internal shells, and some lack a shell.

+ time period: About 540 million years ago to present

+ size: about 0.04–35 in (1 mm–90 cm)

+ where to find: Fine-grained clastic sedimentary rocks and limestones

Gastropods first appeared during the early Cambrian period about 540 million years ago. They are a large and diverse group, with species living in marine environments, freshwater, and terrestrial habitats. Most gastropods have a coiled shell with a single chamber. Unlike bivalves, gastropods have heads that are distinguished from their bodies. Like bivalves, they have a single, muscular foot that helps them move. Gastropod and bivalve fossils are abundant and widespread. Because of their excellent preservation and the diversity of their species, they are important to scientists for understanding evolutionary processes.

Ammonites

Ammonites are an extinct group of mollusks best known for their beautiful spiral shells. Ammonite fossils are prolific in the geologic record.

KEY FACTS

The shells of ammonites are divided into chambers.

+ time period: About 400–65 million years ago

+ size: Up to about 6.5 ft (2 m)

+ where to find: Marine sedimentary rock formations; key localities include Badlands National Park in South Dakota and Guadalupe Mountains National Park in Texas.

Ammonites are marine animals related to squid, cuttlefish, and octopuses, but went extinct about 65 million years ago, the same time as dinosaurs' demise. They are best known for their distinctive spiral shells, though a few species evolved with nonspiral forms. Ammonites were prolific and diverse, with many species. Their fossils are among the most abundant on Earth and serve as excellent time markers, allowing geologists to link the rock layers in which they are found to specific time periods. The name "ammonite" refers to the Egyptian god Ammon, who wore ram's horns resembling the spiral shape of the shells.

Echinoderms

Echinoderms are a diverse group of marine animals including sea stars, sea urchins, and sand dollars. Echinoderms' skeletal plates made of calcite are well preserved as fossils.

KEY FACTS

Fossils of echinoderms show a 5-point radial symmetry.

+ time period: 540 million years ago to present

+ size: About 0.08–8 in (0.2–20 cm); some crinoid stalks are much longer.

+ where to find: Limestones and marine shales

Echinoderms are a broad group of marine organisms that appeared in the early Cambrian period about 540 million years ago, which has many living members today. The most recognizable are sea stars, sea cucumbers, sea urchins, and sand dollars, which have a unique skeleton made of interlocking plates of calcite. These interior, or endoskeletons, called tests, enclose the organs and are covered with a spiny skin. The skeletons also display a radial symmetry, often with five points, such as the five limbs of a sea star. Echinoderm skeletons are widespread in the fossil record and are an important indicator of past marine conditions.

Crinoids

Crinoids are marine animals that look like flowers, with a flexible stalk topped with a head of waving arms. They are echinoderms, related to sea stars and sea urchins.

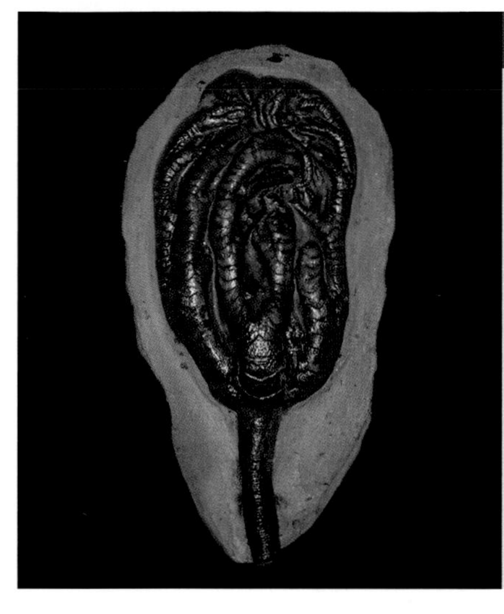

KEY FACTS

Fragmented stems look like small, circular plates; complete fossils are rare.

+ time period: 488 million years ago to present

+ size: Up to about 130 ft (40 m)

+ where to find: Limestones and marine shales; excellent localities are early Devonian sandstones in West Virginia.

Crinoids, like all echinoderms, have a skeleton interlocking calcite plates that house their vital organs. Crinoids are sessile—meaning that they live attached to some sort of substrate such as a rock, shell, or reef—or they are free-floating, planktonic organisms. Crinoids have three basic body parts: a stem, a cup or calyx, and arms. The cup and arms look like tassels or blossoms atop stems. Crinoid stems are flexible cylinders made of small, circular plates with a star-shaped interior canal. Fragments of stems are common fossils, especially in limestones in the western U.S. and Canada. Some limestone formations consist entirely of crinoid parts.

Fish

Animals classified as fish include modern bony fish, sharks, and rays, as well as many species that are extinct. Some mudstones and shales contain exquisitely well-preserved fish fossils.

KEY FACTS

Most fish have elongate bodies with fins, and are covered with scales or bony plates.

+ time period: 500 million years ago to present

+ size: Variable; some fossil fish reach over 65 ft (20 m) long.

+ where to find: Fine-grained sedimentary rocks; famous locales include Fossil Butte National Monument in Wyoming.

Fish are the first vertebrates in the fossil record, about 500 million years ago. Those fish were jawless, and bony plates covered all or most of their bodies, instead of a hard internal skeleton. Over time, different fish appeared, including sharks and rays about 400 million years ago, and eventually fish with internal bony skeletons about 200 million years ago. The Niobrara Formation in the central and western U.S. contains abundant fish fossils from an interior seaway that existed about 85 million years ago. Wyoming's Green River Formation, about 48 million years old, contains spectacular beds of fossil fish.

Shark Teeth

Shark teeth are relatively common fossils and a favorite of collectors. Sharks shed thousands of replaceable teeth during their lifetime—an abundant source for fossilization in ocean sediments.

KEY FACTS

Individual teeth with various shapes can be found.

+ **time period:** 450 million years ago to present
+ **size:** Up to about 7 in (18 cm)
+ **where to find:** Along beaches and eroding out of marine sedimentary rock formations, even far inland such as in many localities in Wyoming

Sharks are among a group of fish that includes skates and rays. They first appeared about 450 million years ago and are distinguished by an internal skeleton of cartilage, not bone. Without hard parts, sharks are rarely preserved as fossils, except their teeth. Most teeth become fossilized by minerals filling in the pores around the tooth material. This process takes thousands of years and is why many fossilized shark teeth are dark instead of the light whites and yellows of living sharks' teeth. A famous fossilized shark is the giant Megalodon, which lived approximately 20 million to 2 million years ago and whose teeth are up to about 7 inches (18 cm) long.

Insects

Insects are among the most diverse group of animals. Wingless insects first appeared about 400 million years ago, and flying insects appeared about 360 million years ago.

KEY FACTS

Wings with a network of veins are more often fossilized than other body parts; whole insects are found in amber.

+ time period: **About 400 million years ago to present**

+ size: **Up to about 2 ft (0.6 m)**

+ where to find: **Fine-grained sedimentary rocks; famous locales include Florissant Fossil Beds in Colorado.**

Insects, one of the most diverse groups of animals on the planet, were the first to develop flight, approximately 360 million years ago. Insects have segmented bodies: a head, a middle section called a thorax, and an abdomen. They have a hard outer skeleton, and their legs are jointed. Insects were particularly abundant about 305 to 270 million years ago, when gigantic dragonflies had wingspans over 2 feet (0.6 m) long. Insects are relatively poorly preserved as fossils because of their fragile bodies. One exception is amber, or fossilized tree resin, which preserves entire insects that became trapped in tree sap.

Amphibians & Reptiles

Amphibians and reptiles are among the earliest animals to inhabit the land. These groups evolved to include a wide diversity of species, both living and extinct.

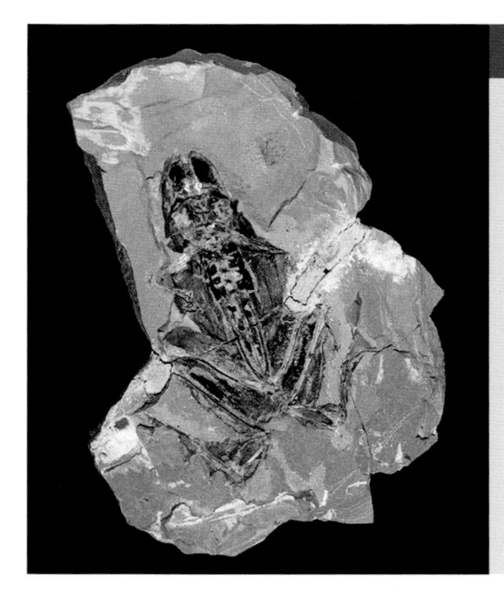

KEY FACTS

Tetrapods, named from the Greek words for "four-footed," are four-limbed vertebrates, and the earliest of these ancient animals able to walk on land were amphibians and reptiles.

+ time period: 368 million years ago to present

+ size: Up to about 130 ft (40 m) long

+ where to find: Fine-grained sedimentary rocks

The first vertebrates to "migrate" from water to the land were amphibians known as the early tetrapods, or four-legged animals. A famous fossil is a 360-million-year-old skeleton of a large creature called Ichthyostega found in Greenland. It more resembled crocodiles than today's amphibians. Others include the bizarre Diplocaulus, a giant newt-like creature with a boomerang-shaped skull. Fossils resembling modern frogs, toads, newts, and salamanders are found in rocks younger than 200 million years old. Reptiles evolved later, and their abundant fossils include turtle shells, lizard skeletons, dinosaur bones, and reptile eggs.

Dinosaurs

Dinosaurs are part of the diapsid group of reptiles. They grew to enormous proportions and dominated most terrestrial habitats until their sudden demise about 65 million years ago.

KEY FACTS

Large bones are found in Triassic to Cretaceous period rocks; dinosaurs had two skull openings behind their eye sockets.

+ time period: 225–65 million years ago

+ size: Variable; up to about 190 ft (58 m) long

+ where to find: Sedimentary rocks; Dinosaur National Monument in Colorado and Utah is a famous site.

Dinosaurs are diverse reptiles that populated the Earth for 160 million years. There are two main groups: Ornithischia, "bird-hipped," with rear-facing pubic bones, and Saurischia, "reptile-hipped," with forward-facing pubic bones. Ornithischia include Triceratops and Stegosaurus. Saurischia include herbivorous sauropods such as Apatosaurus and carnivorous theropods such as Tyrannosaurus and Velociraptor. Paleontologists trace the ancestry of birds to theropods. Recent discoveries indicate that some dinosaurs had extensive feathered plumage. Dinosaur fossils range from footprints to fragments of bone to entire skeletons.

Birds

Bird fossils are of great interest to scientists tracing the evolutionary history from dinosaurs to modern birds. Bird fossils, however, are uncommon because of their fragile hollow bones.

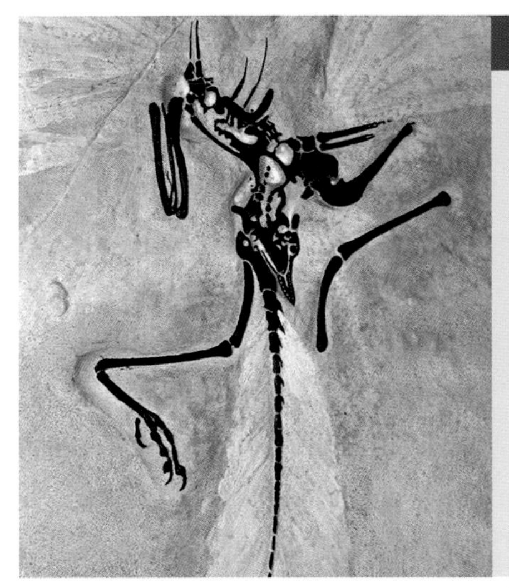

KEY FACTS

Bird fossils are uncommon; evidence for feathers is an important feature.

+ time period: About 150 million years ago to present

+ size: Variable

+ where to find: Sedimentary rock formations; uncommon

One of the oldest and most famous early bird fossils is Archaeopteryx, from a 150-million-year-old limestone quarry in Germany. Archaeopteryx has features of both dinosaurs and birds and has been interpreted as an intermediate evolutionary stage. Like modern birds, Archaeopteryx had wings and feathers, but like dinosaurs, it had teeth, claws, and a long tail. As more discoveries of feathered dinosaurs emerge, however, interpretations of Archaeopteryx and the transition from dinosaurs to modern birds are changing. Birds continued to diversify after non-avian dinosaurs' extinction, and by about 35 million years ago, most groups of modern birds had appeared.

Mammals

The 65-million-year period after dinosaurs' extinction is often considered the Age of Mammals. Their fossils help scientists understand climate changes and movements of the continents.

KEY FACTS

Mammal fossils range from bone fragments and teeth to entire skeletons and even hair.

+ time period: About 150 million years ago to present

+ size: Variable

+ where to find: Widespread; famous localities include La Brea Tar Pits in Los Angeles, Mammoth Site in South Dakota, and sites in western Nebraska.

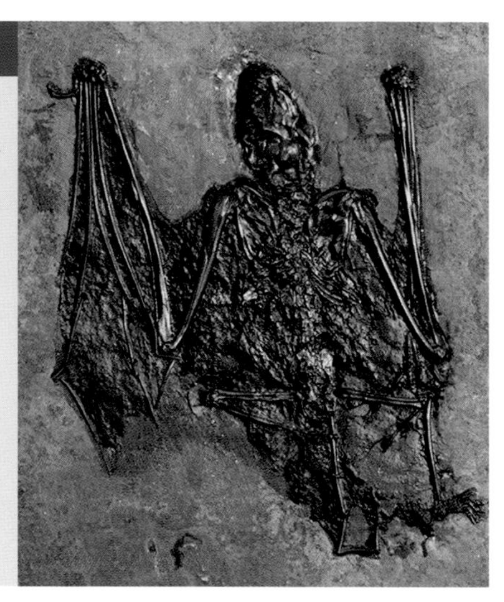

Mammals are warm-blooded, have hair, and feed milk to their young. They appeared around the same time as dinosaurs, but mammals didn't flourish until after dinosaurs' demise about 65 million years ago. Early mammals were relatively small, looked similar to rodents, and fed on insects. As mammals diversified, they became larger and dominated nearly every terrestrial ecosystem. Mammal fossils, especially teeth, are widespread and well preserved. Mammals in famous fossil finds from the Pleistocene, the period of the last ice ages, include skeletons of mammoths, mastodons, saber-toothed cats, native horses, and camels.

Trace Fossils

Trace fossils include footprints, tracks, burrows, bite marks—even fossilized dung. These are traces of living organisms that have been preserved in the geologic record.

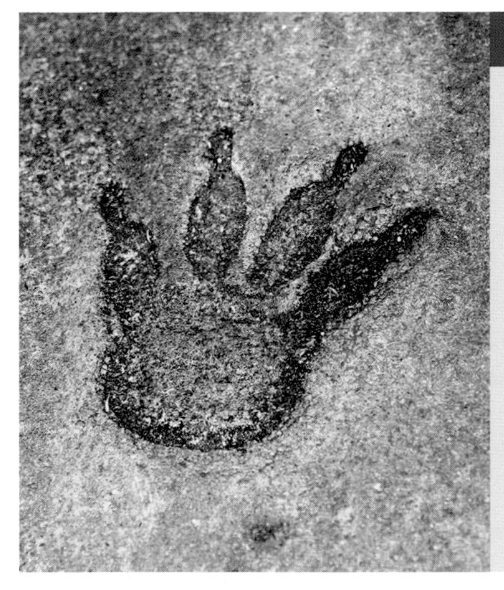

KEY FACTS

Burrow marks, footprints, feeding marks, and excrement are preserved primarily in sedimentary rocks.

+ time period: **About 542 million to 10,000 years ago**

+ size: **Variable**

+ where to find: **Sedimentary rocks; locales include dinosaur tracks at Dinosaur State Park in Connecticut and Dinosaur Valley State Park in Texas.**

Trace fossils preserve organisms' activities rather than their body parts. Invertebrates' trace fossils include impressions of feeding, burrowing, and boring, mostly preserved in fine-grained marine sedimentary rocks. Trace fossils from vertebrates, especially land animals, include bite marks and footprints, even pieces of fossilized dung known as coprolites. Trace fossils provide an understanding of how prehistoric organisms lived, moved, and ate. Sometimes it is difficult to distinguish a trace fossil from a pseudofossil, a geological formation that looks similar to something made by an organism but is inorganic.

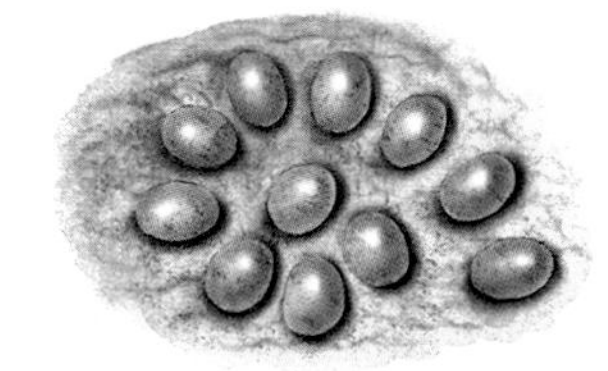

Lineation

Deformed rocks

Lineations are linear structures in deformed rocks. There are several types, including the hinges of folds and the long axes of aligned minerals and stretched clasts.

KEY FACTS

Lineations are structures that can be represented by a line: long in one dimension relative to others.

+ fact: Look for lineation along the surface of a foliation plane.

+ fact: Deformed rocks with strong linear fabrics and without planar fabrics (foliation) are L-tectonites.

+ fact: Common in mountain outcrops and ancient cores of continents

A lineation is a visible texture of linear elements that develops when a rock is deformed. For example, a conglomerate originally made of rounded cobbles develops a lineation if it is squeezed such that the cobbles are stretched out into elongate, cigar-like shapes (it develops a foliation if the clasts are flattened into pancakes). Other common lineations are defined by the alignment of elongate minerals, which can form as metamorphic minerals grow into a preferred orientation, or they can be original minerals that are rotated into a linear fabric during deformation.

Foliation

Deformed rocks

Deformed rocks that exhibit layering or any through-going textures that are planar are said to have foliation. These rocks may split or break along foliation planes, or may be more massive.

KEY FACTS

Foliation is an arrangement of planar or tabular features in a deformed rock.

+ fact: Rocks can have more than one type of foliation in more than one orientation.

+ fact: Layering in schists and gneisses is an easy way to identify foliation.

+ fact: Common in outcrops in mountain belts and the ancient cores of continents

A rock has a foliation if it has a visible texture or fabric created by planar or tabular elements. These form as a rock is squeezed and flattened by forces in the Earth's crust that are typical of mountain-building processes. Slates are metamorphic rocks that cleave or break along foliation planes. The fine layering in schists created by aligned plates of mica is another type. In higher-grade metamorphic rocks such as gneisses, light and dark layers of different minerals define a foliation. Sometimes rocks display more than one foliation.

Augen

Deformed rocks

Augen are large, lens-shaped mineral grains found in deformed rocks. Augen are typically surrounded by fine grains of mica and other minerals aligned in wavy layering or foliation.

KEY FACTS

Eye-shaped minerals, commonly feldspars, are surrounded by platy minerals aligned in a foliation.

+ fact: Augen are larger than other mineral grains in a rock, commonly about 0.2–1.2 in (0.5–3 cm).

+ fact: Found in metamorphic rocks such as schists and gneisses

+ fact: Form when rocks are deformed at depth in the Earth's crust

Augen, German for "eyes," are eye-shaped minerals or clusters of minerals in some deformed rocks. They can be relict crystals from the original rock that were rotated and rounded during high-temperature deformation, or they can be newly grown metamorphic crystals that developed during deformation. Augen are typically harder, compact minerals such as feldspar, quartz, and garnet. They are surrounded by micas and other platy crystals that are aligned in a wavy foliation. Augen form in areas of the Earth's crust called shear zones, where rocks move past each other at relatively high temperatures.

Boudinage

Deformed rocks

Boudinage forms when a relatively rigid rock layer is stretched and pulled apart into sausage shapes known as boudins. Surrounding layers appear to flow around the boudins.

KEY FACTS

A rock layer or quartz vein looks as if it has been pulled apart like taffy.

+ **fact:** Easiest to see in exposures perpendicular to the rock's dominant foliation or layering

+ **fact:** Found in deformed veins and foliated rocks

+ **fact:** Forms when rocks are deformed in the Earth's crust before they are exposed at the surface

Boudinage is a structure in veins or rock layers that are more rigid or competent than the surrounding rock. When these are stretched, the competent vein or layer is pulled apart. The result is an effect that resembles a link of sausages. Boudinage is from the French *boudin,* a type of sausage. Boudinage structures range from thin, undulating ribbons in narrow veins, to larger, blocky, rotated tablets. Boudinage is an example of how different rocks react to the same forces: Rock around boudins behaves like heated plastic, flowing in a solid state to accommodate the strain, while the mechanically stronger boudin layer pulls apart like taffy.

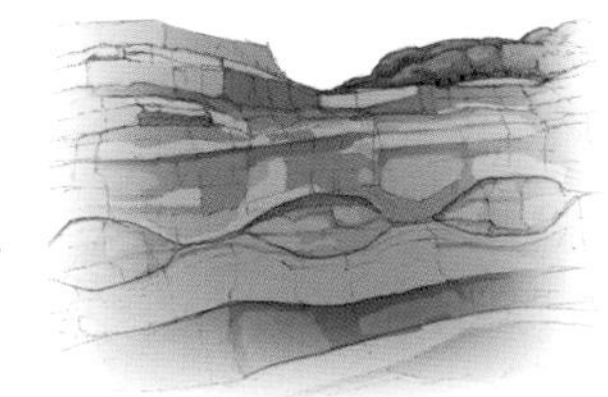

Folds

Deformed rocks

Folds are bends in bodies of rock that form during shortening or contraction. Folds occur at all scales, from deformation of individual crystals to structures the size of mountain ranges.

KEY FACTS

This deformation refers to folded, bent, or crenulated features in a rock mass.

+ fact: Found in all types of rock

+ fact: Small folds in an outcrop can mirror larger folds on a landform scale.

+ fact: Large folds create parallel ridges and valleys in the Appalachian region of the U.S.

Folds are formed when rocks shorten or contract due to movements in the Earth's crust and upper mantle. Folds can be found at all scales, from bent individual mineral grains to enormous warps defining mountain ranges. Geologists have many terms to describe different kinds of folds. Folds in the shape of a lowercase *n* are called anticlines; folds in the shape of a lowercase *u* are called synclines. Folds can be upright, asymmetrical, or recumbent. Tight, Z-shaped folds are called crenulations or chevrons. Folded sedimentary rocks can form traps for accumulation of oil and natural gas.

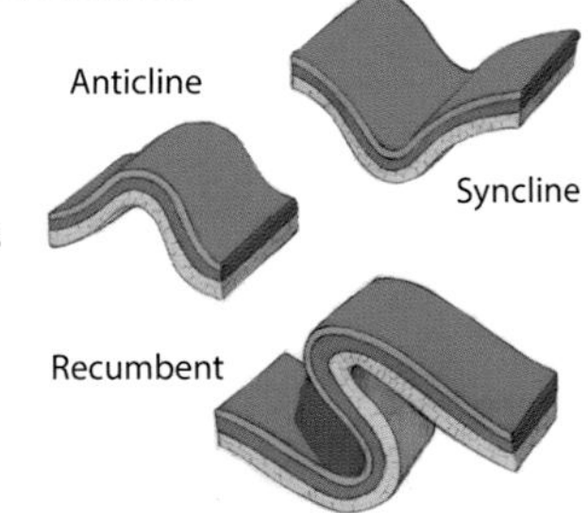

Faults

Deformed rocks

Faults are brittle fractures that accommodate movement in rocks. The body of rock on one side of a fault moves relative to the body on the other side by sliding along the fault plane.

KEY FACTS

In faults, movement is apparent by offsets or juxtaposition of different rock formations.

+ fact: Movement, or slip, releases energy as earthquakes.

+ fact: Faults are rarely single, clean fractures; they are typically networks of fractures and broken rock.

+ fact: Active faults are common along boundaries of the Earth's plates.

Three types of faults are thrust or reverse, normal, and strike-slip. Thrust or reverse faults are planes along which deeper rocks move upward relative to shallower rocks. These faults are caused by compressional forces in the crust. Normal faults are planes along which younger, shallower rocks drop down relative to older, deeper rocks. Normal faults accommodate stretching or thinning in the crust and are common in the U.S. Basin and Range province. Strike-slip faults are planes along which rocks slide laterally past each other without significant vertical movement. The San Andreas Fault is a famous strike-slip fault.

Joints

Deformed rocks

Joints are naturally occurring cracks found in almost all rock types. Joints are distinguished from faults by a lack of movement across the fracture plane.

KEY FACTS

To see a joint, look for brittle cracks that cut across rock faces.

+ fact: Joints form from stresses in the Earth's crust.

+ fact: Joints are important conduits for groundwater in aquifers.

+ fact: Joints can have various orientations, appearing in regular sets or as isolated fractures.

Many rock outcrops contain brittle cracks, called joints. There is no movement along joints, however. If there is evidence of sliding along a fracture, the crack is called a fault, not a joint. Joints form by forces that act upon a body of rock. As the Earth's plates move, forces are distributed through the crust, causing some rocks to crack. Rocks can also crack during changes in temperature, from uplift to shallower depths in the Earth's crust, or by cooling of magma. Joints occur as isolated cracks or in widespread fractures. They are affected by weathering and erosion, and are an important control on the evolution of a landscape.

Veins

Deformed rocks

Veins are narrow, sheetlike bodies made up of one or more minerals. Veins form when minerals crystallize out of water-rich fluids.

KEY FACTS

Veins appear as crosscutting stripes or stringers on the surface of an outcrop.

+ fact: The most common veins are white and are filled with quartz or calcite.

+ fact: Veins can also be host to rare minerals and ore deposits.

+ fact: Veins are found in all rock types.

Cutting across many outcrops are narrow, sheetlike bodies made up of one or more minerals. These are called veins, and they are filled with minerals that crystallized out of water-rich fluids. In some cases, veins form when fluids infiltrate existing cracks or joints in a rock formation. Veins following crack systems may form individual tabular bodies or spidery networks of irregular shapes. Some veins are tightly folded, indicating high-temperature deformation of the rock body after the vein crystallized. Veins are important sources of metals and other precious minerals, including gold and copper.

Tafoni

Weathering & erosion

Tafoni, or honeycombing, is a surface feature marked by rounded pits, hollows, and shallow caverns, often clustered in groups and separated by hardened ridges or visors.

KEY FACTS

These features are easiest to identify in cliff faces with a honeycomb appearance.

+ fact: Develop in many different rock types

+ fact: Appear as small pits clustered together or large, rounded hollows

+ fact: Found in sandstone cliff faces and some granitic bodies; common in coastal environments

Clusters of hollows, rounded pits, and shallow caverns on rock faces and outcrops are known as tafoni, honeycombing, or cavernous weathering. Geologists have posed many hypotheses for how cavernous weathering develops. Most agree that salt crystallization is key. Salts crystallize in tiny pits on the surface of a rock, transported by wind or water. The salt expands, breaking off adjacent grains of the host rock and expanding the pit. This process forms larger and larger rounded hollows. In some porous sandstones, cement hardening along internal joints or bedding planes may be important in channeling erosion.

Desert Varnish

Weathering & erosion

Desert varnish is a surface coating common on the sandstone rock walls of the desert Southwest, though it can form on many rock types and in different environments.

KEY FACTS

The varnish is a crust on the surface of a rock; dark colors vary from reddish brown to shiny black.

+ fact: Crust is usually less than 0.02 in (0.5 mm) thick.

+ fact: Common on exposed rock faces in arid environments

+ fact: Many petroglyphs are created by chipping through desert varnish to paler rock beneath.

Desert varnish is a dark, hard coating made of clay minerals, manganese oxide, and iron oxide. It is found on many different rock exposures. Geologists have debated the varnish's formation because manganese and iron oxides could indicate bacterial metabolism. Rocks on Mars appear to have desert varnish; however, most geologists agree that it can develop without the aid of life. Much of the varnish is composed of silica, which reaches the surface via clay minerals in windblown dust or by leaching from the rock's interior. Black, shiny desert varnish is rich in manganese oxide; dull reddish varnish is rich in iron oxide.

Spheroidal Weathering

Weathering & erosion

Spheroidal weathering, also known as onion-skin weathering, is a process that creates rounded outcrop formations and boulders in the shape of spheres.

KEY FACTS

Exposed rounded stones may be with or without surrounding disintegrating crusts.

+ fact: Can occur in many different rock types

+ fact: Rounded products of spheroidal weathering are called "core stones."

+ fact: In granite, the disintegrating rock around core stones is called "grus."

Spheroidal weathering is a type of weathering that results in rounded outcrop formations and spherical boulders. Spheroidal weathering occurs when a rock formation is broken into blocks by several sets of fractures called joints. Groundwater penetrates the rock formation along the joints, facilitating chemical weathering at a faster rate than where the rock is not fractured. Physical weathering is also accelerated along joints. Weathering progresses, rounding off the edges of two intersecting joints and the corners of three intersecting joints, producing curved rock formations and even perfect spheres.

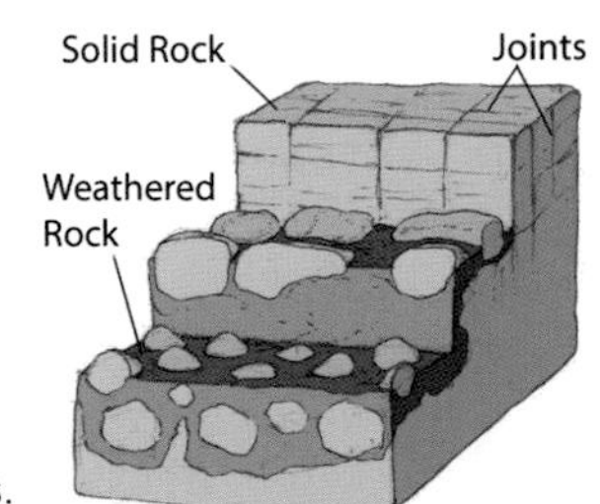

Sheeting or Exfoliation Joints

Weathering & erosion

Sheeting joints, also known as exfoliation joints, are curved fractures that parallel the surface of a body of rock. The joints separate sheets of rock, creating dome-like landforms.

KEY FACTS

These joints are most common in granite but can form in other rock types.

+ fact: Sheeting joints separate slabs of rock of various thickness.

+ fact: Weathering and erosion along sheeting joints create spectacular landforms such as Half Dome in Yosemite National Park.

+ fact: Rockfall can occur when slabs fail along the joints.

Sheeting joints, also known as exfoliation joints, are common features of granitic terrains. These are sets of curved fractures oriented roughly parallel to the surface of a rock. The joints separate concentric slabs or shells of rock that progressively weather and erode, creating domes and other rounded landforms. One explanation for sheeting joints is that they form during the release of pressure when rocks overlying a body of granite are removed by erosion—a removal called "unroofing." Another explanation is that sheeting joints form from mechanical stresses along the surface of a dome.

Arches

Weathering & erosion

Natural arches, windows, and bridges are erosional features of some landscapes. They can form by river erosion, wave action along coastlines, and weathering.

KEY FACTS

Natural stone arches, windows, or bridges are spectacular features of eroded landscapes.

+ fact: Form mostly in sedimentary rocks like sandstones and limestones

+ fact: Arches are formed by erosion.

+ prime locations: Arches National Park and Zion National Park in Utah; Natural Bridges National Monument in Kentucky

Arches, windows, and bridges are slabs of rock overlying natural openings. Many arches are formed by erosion along parallel sets of fractures to create rock fins. Some fins are undermined when there is contact between a permeable sandstone formation and an impermeable formation such as shale. Groundwater percolating through the sandstone stops at the shale and flows laterally, weakening the sandstone. This can undermine a sandstone fin, causing a rockfall or opening an arch. Natural bridges and windows may also form by dissolution of limestone in cave formation.

Three stages in arch formation

Towers & Fins

Weathering & erosion

Along with arches, natural stone towers and fins are some of the world's most spectacular landforms. Most of these features form by preferential weathering and erosion along fractures.

KEY FACTS

Stone towers, needles, and fins rise above the land surface.

+ fact: Several processes can create these formations.

+ fact: Towers form in many rock types; common in sedimentary and volcanic rocks.

+ prime locations: Canyonlands National Park, Castle Valley, and Monument Valley, all in Utah

Towers and fins are iconic landforms of the Southwest. Most of them have been created by millions of years of weathering and erosion. The key to the creation of towers and fins is in the regional pattern of fractures in the rock, known by geologists as joints. Weathering and erosion become concentrated along these planes, and if a rock contains a set of evenly spaced, parallel fractures, erosion along these planes will leave narrow fins standing. If a rock contains two sets of joints that are oriented roughly perpendicular to each other, weathering and erosion will create freestanding towers.

Hoodoos/Mushroom Rocks/Pedestal Stones

Weathering & erosion

Hoodoos, mushroom rocks, and pedestal stones are names given to rock spires and perched blocks with irregular, sometimes fantastical shapes.

KEY FACTS

The rock spires may or may not have capstones or perched blocks.

+ fact: Common in sedimentary and volcanic rocks

+ fact: Found especially in deserts, badlands, and mountains

+ prime locations: Bryce Canyon National Park in Utah, Badlands National Park in South Dakota, and Tent Rocks National Monument in New Mexico

Thin rock spires with irregular shapes are known as hoodoos, tents, and fairy chimneys. They are found in basins, canyons, and badlands, usually in layered sedimentary or volcanic rocks. Spires supporting perched blocks or wide, rounded caps are known as mushroom rocks or pedestal stones. These are commonly in granitic, volcanic, and sedimentary rocks. Many hoodoos and pedestal stones form when a hard, more resistant rock formation overlies a softer, weaker formation. Weathering attacks these rocks along joints, leaving narrow cones of the softer rock protected by isolated caps of the overlying formation.

Three stages in formation of hoodoos

Landslides/Slumps/Creep

Weathering & erosion

The movement of rock and soil down a slope creates distinctive land features and gravity-formed deposits. Downslope movement occurs via slides, slumps, flows, or slow creep.

KEY FACTS

Indicators of downslope movement include curved trees, rock scars, scarps, and broken or hummocky land surfaces.

+ fact: Occurs when slopes are destabilized by natural factors and/or human activities

+ fact: Landslides are a major hazard.

+ fact: Slides move as slowly as a few millimeters a year or up to 200 mph (322 kph).

Soil or rock creep is a slow drift of material down a slope. Its indicators can be subtle. Vegetation may continue to grow upward on a creeping slope, resulting in trees with curved trunks. Material in a slide mainly retains its structure and catastrophically fails all at once, usually on underground surfaces lubricated with water. Slumps occur when groundwater disrupts the structure of a body sliding down a curved surface. Flows, including slow creep, occur when material is saturated with water and behaves like a viscous liquid, as in mud and debris flows.

Dikes

Igneous terrain

Dikes are sheets of igneous rock that crosscut surrounding bedrock at a steep angle. Basalt dikes commonly occur in areas of extension in the Earth's crust.

KEY FACTS

This band or sheet of rock crosscuts surrounding rock at a steep angle.

+ fact: Dikes are made of igneous rock, but they can crosscut any rock type.

+ fact: Large numbers of dikes in a region are called dike swarms.

+ fact: Widespread; common in volcanic terrain, granitic regions, and mountain belts

Dikes are igneous intrusions that crosscut bedrock as steeply angled or vertical sheets. Different types of dikes form in different magmatic environments. Dikes form when magma rises in the Earth's crust until reaching neutral buoyancy—the same buoyancy as surrounding rock. Then, the magma spreads laterally, vertically, or horizontally depending on local stresses and weaknesses. Vertical orientations create dikes; lateral orientations create sills. Dikes with finer grain sizes along their margins than in their interior are called chilled margins. These form when magma cools more quickly in contact with surrounding rock.

Sills

Igneous terrain

Sills are intrusive sheets of igneous rock oriented horizontally or parallel to the fabric of the surrounding bedrock. They are common in sedimentary basins intruded by magma.

KEY FACTS

This band or sheet of igneous rock parallels the structure of the surrounding bedrock.

+ fact: Sills are commonly thicker than dikes.

+ fact: Sills consist of igneous rock but can intrude any rock type.

+ fact: Widespread; common in volcanic terrain and regions of sedimentary bedrock

Sills are sheetlike intrusive bodies oriented parallel to the layering of surrounding rock, commonly horizontal in flat sedimentary basins. Sills are like dikes, forming from lenses of buoyant magma that spread laterally in the crust. In sills, stresses in the Earth's crust favor emplacement horizontally or parallel to existing structures. These generally form near the Earth's surface and can be thicker than dikes because pressures tend to be lower closer to the surface. The Palisades Sill in New York and New Jersey is a diabase sill that forms spectacular cliffs along the Hudson River.

Inclusions

Igneous terrain

Some intrusive igneous rocks contain inclusions of other rocks. Inclusions can result from magma or from blocks of surrounding rock entrained in magma as it cooled.

KEY FACTS

Inclusions are angular blocks or rounded blobs of rock that are compositionally different from the surrounding rock.

+ fact: Inclusions are found in both intrusive and extrusive igneous rocks.

+ fact: Some inclusions are called xenoliths, from the Greek word *xeno*, meaning "foreign."

+ fact: Some xenoliths contain diamonds.

Large exposures of intrusive igneous rocks such as granites are heterogeneous. Granites often show variability in grain size, crystal orientation, and mineralogical makeup. Some granites (and other igneous bodies) contain inclusions of other rock types. Inclusions can be chunks of mantle rock brought to the surface by ascending magma. These inclusions, called xenoliths, can tell us about rocks deep in the Earth's interior. Inclusions of basalt and/or diorite are also common in granitic bodies. These are thought to be blobs of mafic magma that became entrained when the granite was not fully hardened.

Volcanic Bombs

Igneous terrain

Volcanic bombs are pieces of lava ejected during volcanic eruptions. These masses solidify in the air before landing and can have shapes like teardrops, ribbons, or balls.

KEY FACTS

Chunks of volcanic rock have distinctive shapes including teardrops, ribbons, and spindles.

+ fact: Volcanic bombs are 2.5 in (6.4 cm) to many feet in diameter.

+ fact: Can form out of different types of lava; basaltic bombs are common.

+ fact: Occur on flanks of volcanoes such as Mauna Kea in Hawaii

Volcanic bombs are ejected chunks of magma that form streamlined masses as they harden while flying through the air. Volcanic bombs are larger than about 2.5 inches (6.4 cm) in diameter. Smaller particles are called lapilli or ash. Bombs commonly form elongate spheroidal bodies, sometimes with twisted "tails" or spindle shapes. Some are not completely solidified when they hit the surface, and they flatten upon impact. Other shapes include teardrops and ribbons. Volcanic bombs are ejected along vents that liberate gases as large, bursting bubbles and indicate more explosive varieties of volcanoes.

A'a Lava

Igneous terrain

A'a (pronounced ah-ah) is a Hawaiian word for a type of lava flow characterized by sharp, angular rubble. It forms when the top parts of a flow harden and break into chunks.

KEY FACTS

These lava flows produce piles of sharp, rubbly volcanic rock.

+ fact: Develops where lava flows over steep slopes

+ fact: Flows typically advance faster than pahoehoe flows.

+ fact: A'a is common on the Hawaiian Islands.

Basaltic lava flows have three types: a'a, pahoehoe, and pillow lava. A'a creates piles of sharp, jagged rubble. A'a forms when the surface of a flow hardens into a solid and then breaks up into fragments because of the continued movement of lava beneath. The fragments tumble down the front of the flow and are entrained beneath it as molten lava in the interior continues to advance. This movement is similar to the tread of an advancing bulldozer. A'a flows create thick piles of sharp fragments and blocks that are difficult to walk on. Chunks of a'a are black or red volcanic rocks often used in landscaping.

Pahoehoe Lava

Igneous terrain

Pahoehoe (pronounced pa-hoy-hoy) is a Hawaiian word for a type of lava flow characterized by smooth surfaces with wrinkles and ropelike ridges.

KEY FACTS

Pahoehoe forms hardened lava flows with smooth surfaces and curved ridges.

+ fact: Pahoehoe can transition to a'a, but a'a does not turn into pahoehoe.

+ fact: Footpaths on volcanic rock often follow pahoehoe flows.

+ fact: Pahoehoe is common on the Hawaiian Islands.

Pahoehoe, unlike a'a, is a smooth and cohesive lava flow. It forms when the surface of molten lava cools into a thin, smooth skin. The skin is then pushed into folds by faster-moving lava below. Pahoehoe flows can eventually develop into lava tubes. When the flows harden, accumulations of the surface folds create wrinkles and ropelike masses. The surface of pahoehoe is relatively smooth, and footpaths on volcanic rock usually follow exposures of pahoehoe. Basaltic volcanoes often have both pahoehoe and a'a lava flows, depending on the viscosity and temperature of the lava as well as on the shape of the terrain.

Pillow Lava

Igneous terrain

Pillow lava is a type of lava flow that forms underwater. Outcrops look like stacks of elongate lobes and rounded blobs. Exposures of ancient pillow lavas can be highly deformed.

KEY FACTS

Formations consist of stacked lobes and protrusions, circular or elliptical in cross section.

+ fact: Outer layer is typically smooth and can be glassy.

+ fact: Pillow lavas in ancient rocks indicate the presence of water.

+ fact: Active pillow lavas are under Hawaiian waters; ancient pillow lavas are in uplifted oceanic rocks.

Pillow lava is formed when lava erupts underwater, along mid-ocean ridges and submarine flanks of seamounts and other volcanoes. Pillow lava is named for the globular shape of lava as it cools underwater. It develops because the surface of the lava rapidly cools and hardens to form a crust. As the molten lava continues to advance, it breaks through the crust in the shape of rounded blobs or tubes that are hard outside and molten inside. These blobs pinch off from the crust and roll down to the front of the flow. Pillow lavas in exposed ancient rocks are evidence of past oceanic environments.

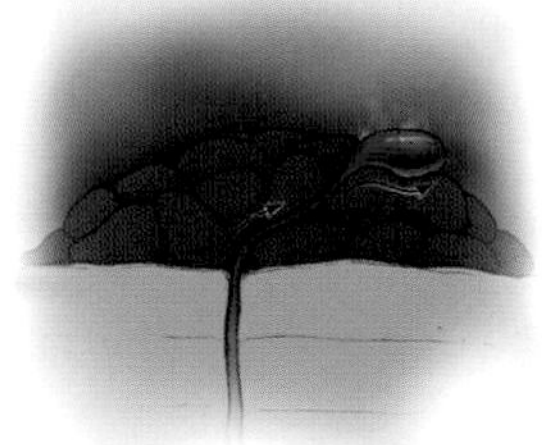

Lava Tubes

Igneous terrain

Lava tubes are open tunnels made of shells of volcanic rock, typically basalt. They are conduits for molten lava. Some volcanoes have networks of lava tubes along their flanks.

KEY FACTS

Open tunnels are formed by hardened volcanic rock, commonly basalt.

+ fact: Lava tubes form in flows of low-viscosity lava.

+ fact: Lava tubes can create intricate networks of caves.

+ fact: Well-known lava tubes are found in New Mexico, Hawaii, California, Oregon, and Washington.

Lava tubes are large hollow tunnels through which lava travels away from its vent. Lava tubes tend to form in volcanoes that have relatively low eruption rates. High eruption rates can sustain open channels rather than closed tubes. Lava tubes are created when the surface of a moving lava flow cools and hardens into a crust. The crust insulates the lava flow below from the cool atmosphere, allowing it to remain molten. These streams of molten lava typically continue to drain through the tube until the magma supply is cut off. Usually the lava then drains away, leaving an empty tube in the volcanic landscape.

Cinder Cones

Igneous terrain

Cinder cones are small volcanoes that form out of hardened pieces of ejected lava that rain back down from the atmosphere after they are erupted.

KEY FACTS

These cone-shaped hills are made of chunks of hardened lava.

+ **fact:** Typically less than 1,000 ft (300 m) high

+ **fact:** They are built up over the course of multiple eruptions over many years.

+ **fact:** Examples are in Sunset Crater Volcano National Monument in Arizona and Lava Beds National Monument in California.

Cinder cones are a simple type of volcano in which most of the erupted magma is ejected as a fountain-like spray out of a central vent. The magma feeding cinder cones is charged with gasses. These gasses expand as they rise, causing the lava to spray out of the vent. The airborne lava hardens into volcanic bombs and smaller chunks of volcanic rock called cinders, lapilli, and ash. These particles, collectively known as tephra, fall back to Earth and accumulate around the vent, building up a cone-shaped landform. Most cinder cones have a rounded crater at their summit.

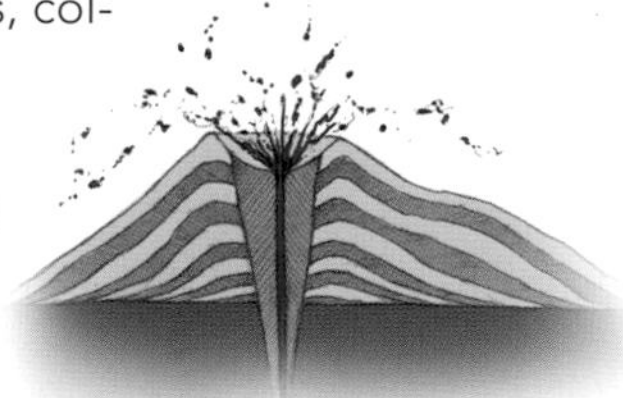

Caldera

Igneous terrain

A caldera is a large depression of rock formed in a volcano. Calderas form when volcanic eruptions empty underground magma chambers, causing the rock above to collapse.

KEY FACTS

These large volcanic depressions are often several miles wide.

+ fact: Common features of explosive volcanoes

+ fact: They're sometimes confused with craters, which are smaller, bowl-like formations at the summit of volcanoes.

+ fact: Much of Yellowstone National Park sits within an enormous caldera.

A caldera, named from the Spanish word for cauldron, is a large depression created by volcanic activity. It forms when a magma chamber beneath a volcano is emptied by eruption, causing the rocks above to collapse into a large depression. In very explosive volcanoes, calderas can form by a single eruption. In less explosive volcanoes, magma chambers may take numerous small eruptions to empty; thus, caldera formation is a slow, progressive process. Continued volcanic activity after formation of a caldera may cause the magma chamber to recharge, and then the center of the caldera will rise into a feature called a resurgent dome.

Volcanic Neck or Plug

Igneous terrain

A volcanic neck is the resistant core of a volcano that is exposed by erosion. Volcanic necks, also called plugs, are typically tall, somewhat cylindrical-shaped landforms.

KEY FACTS

A neck is an eroded cylindrical landform made of volcanic rock.

+ fact: Often found with eroded dikes; some surround the neck radially.

+ fact: Volcanic necks sometimes contain pieces of rock pulled from deep in the Earth during eruption.

+ fact: Famous necks include Devils Tower National Monument in Wyoming and Shiprock in New Mexico.

Most volcanoes are fed by a central vent that pipes magma from deep within the Earth. Conelike outer structures of many volcanoes are created by buildup of lava flows and/or accumulation of ejected ash, cinder, and volcanic bombs. After volcanic activity ends, these are worn away by erosion. Outer flanks of volcanoes are sometimes easily eroded, especially if they consist of unconsolidated cinder and ash. Often the most resistant part of a volcano is the vent that becomes filled with hard volcanic rock at the end of eruption. Remnants of these vents projecting above the landscape are called volcanic necks or plugs.

Columnar Jointing

Igneous terrain

Symmetrical columns of volcanic rock form by a process called columnar jointing. As lava shrinks during cooling, fractures divide the rock into roughly hexagonal columns.

KEY FACTS

Polygonal columns consist of volcanic rock.

+ fact: Form in lava flows, ash flows, and shallow intrusions

+ fact: Columns can be hundreds of feet high.

+ fact: Famous examples are in Devils Postpile National Monument in California and Devils Tower National Monument in Wyoming.

Columnar jointing is a distinctive outcrop feature of some volcanic rocks, especially basalt flows in the West. As thick flows of lava cool, they shrink, causing the hardening lava to crack. In a homogeneous pool of lava without outside stresses, vertical cracks will develop with intersections of approximately 120 degrees, which is the most efficient orientation to accommodate shrinking. This orientation results in vertical columns with hexagonal cross sections. In the real world, however, columnar joints can have variable angles, and the resultant basalt columns take on various polygonal cross sections.

Bedding

Sedimentary terrain

Sedimentary beds are layers of rock with distinguishable tops and bases. Bed boundaries form by changes during deposition of sediment, before the sediment hardens into a rock formation.

KEY FACTS

Layers can be seen in sedimentary rock.

+ fact: Range from the millimeter to tens of meters scale

+ fact: Deposited horizontally with older layers below younger ones

+ fact: Dramatic horizontal patterns in walls of the Grand Canyon are created by bedding planes.

Sedimentary rocks are easily recognized by their layered appearance. The layers, called beds or strata, are made of accumulated sediments compacted and/or cemented into stone. Sediments are transported to their resting place by a mechanism such as wind or water, or they are deposited in place by living organisms such as coral, or by chemical precipitation out of a solution. Changes in this process are caused by changes in the environment or in the sediment source. The changes create natural breaks in the pile of sediment, which later define bed boundaries or bedding planes in the sedimentary rock.

Unconformity

Sedimentary terrain

An unconformity is a break in the sequence of sedimentary beds along which erosion occurred. There are three types: angular unconformities, disconformities, and nonconformities.

KEY FACTS

This feature is caused by disruption in the layering of sedimentary rocks.

+ fact: An unconformity represents a gap in the rock record.

+ fact: Unconformities can indicate deformation, erosion, or changes in sea level.

+ fact: John Wesley Powell described the Great Unconformity in the Grand Canyon in 1869.

In an angular unconformity, sedimentary layers below the boundary are at a different angle from layers above. These occur when the lower rocks are folded and/or faulted so that the layers are no longer horizontal. Later, new sediment is deposited, creating a new horizontal sequence of layers. Disconformities occur when there is a break in sediment deposition and erosion of a surface before the upper layers are deposited. These boundaries often show irregular surfaces between different rock types. A nonconformity is an erosive surface between a nonlayered rock such as granite and a layered sedimentary sequence.

Cross Bedding

Sedimentary terrain

"Cross bedding" is a term for layering that forms within a sedimentary bed and is oriented at an angle to the original horizontal bedding planes of the rock.

KEY FACTS

Angled layers occur within a sedimentary bed.

+ fact: Found in sedimentary rocks, primarily sandstones

+ fact: Some cross beds are a mark of ancient sand dunes.

+ fact: Geologists use cross beds to interpret previous environments and even the direction of water and wind currents.

Cross beds are sublayers in sedimentary rock formations that are oriented at an angle to the bedding planes. Cross beds form from particles moved by water or wind as they are being deposited. The grains accumulate into piles; in windblown sand, the piles can be large dunes. When they grow too high, the grains avalanche down the front and come to rest. Continued movement of particles rebuilds the piles. Repeated cycles create inclined layers or laminations that are preserved in rock. Cross bedding forms dramatic textures in sandstone outcrops in the Southwest, including famous exposures in Zion National Park in Utah.

Ripple Marks

Sedimentary terrain

Ripples are low, elongate ridges that form when water or wind move sand particles. Ancient ripples are preserved in sandstones as features known as ripple marks or cross lamination.

KEY FACTS

Planar surfaces of sedimentary rocks contain sets of low, elongate ridges.

+ fact: Found in sedimentary rocks, primarily sandstones

+ fact: Look for ripple marks in outcrops of sedimentary rock in the Southwest.

+ fact: Geologists use ripple marks to interpret prior environments.

Hiking on sedimentary rocks, you may find exposed bedding planes in sandstones that are marked by low, elongate ridges—semiparallel sets of wrinkles protruding from the surface. These are ripple marks, which are essentially the preserved trace of sand ripples that formed during the deposition of the sand particles that make up the rock. Ripples are small, generally about an inch high, and form by the way sand particles aggregate and move as they are transported by currents or waves. Ripples can form from wind in desert environments or by the flow and/or waves in rivers, lakes, and seas.

Mudcracks

Sedimentary terrain

Mudcracks are the cracks separating polygonal forms in clay and mud beds. Cracks form when wet, fine-grained sediments shrink as they dry.

KEY FACTS

Networks of cracks break dried mud or clay into plates.

+ fact: Ancient mudcracks are preserved in claystones and mudstones.

+ fact: Dinosaur tracks are commonly found in the same rock formations as mudcracks.

+ fact: Look for mudcracks in outcrops in Colorado, New Mexico, Utah, and Arizona.

When a bed of wet, fine-grained sediment such as a layer of clay or mud dries, the surface shrinks and contracts, resulting in a network of cracks. The cracks form plates of sediment with curled edges. The plates can form rough polygonal shapes, commonly hexagons. Environments creating mudcracks include marshes and lake beds that drain and become exposed to the atmosphere. Mudcracks can be preserved in sedimentary rocks if they are filled with sediment and then buried. Geologists use mudcracks in ancient rocks to understand "which way was up" in sedimentary rocks that have been disturbed by folding or faulting.

Karst

Sedimentary terrain

Karst is a distinctive type of terrain formed by the chemical dissolution of rocks. It develops in soluble rocks, primarily carbonates like limestone and dolomite.

KEY FACTS

Karst landforms include caves, springs, towers, rocks in fluted shapes, stalactites, and stalagmites.

+ fact: About 10 percent of the Earth's surface is karst.

+ fact: Much of the world's life depends on water from karst areas.

+ fact: Kentucky's Mammoth Cave is a famous karst, and karst is actively forming in Florida.

Carbonate rocks, especially limestone, dissolve in contact with groundwater or seawater that is undersaturated with calcium carbonate. Rock formations made of evaporite deposits such as gypsum and halite will also dissolve in contact with water. Landforms that develop by dissolution of bedrock are called karst. Karst surfaces are marked by springs, sinkholes, and caves. Dissolution along joints and faults can result in karst towers with fluted surfaces. Formations range from tiny chemical precipitates to entire landscapes. Water in underground conduits and caves, called karst aquifers, are important water resources.

Stalactites & Stalagmites

Sedimentary terrain

Stalactites and stalagmites are features of limestone caves. They are a type of cave deposit called dripstones that are formed by drops of water that precipitate calcium carbonate.

KEY FACTS

Stalactites point down from roofs of caves; stalagmites point up from floors.

+ fact: Stalactites and stalagmites are karst formations.

+ fact: Most are made of calcium carbonate.

+ fact: Spectacular examples include Natural Bridges Caverns in Texas, Carlsbad Caverns National Park in New Mexico, and California Caverns.

Stalactites and stalagmites are cave deposits called speleothems. The word "speleothem" is from the Greek *spelaion,* meaning "cave," and *thema,* meaning "deposit." Stalactites are cone or straw shaped, and protrude downward from the roof. These form from water trickling down from cracks in the cave roof. Stalagmites are cone shaped and protrude upward from the floor; they form from drops of water falling from stalactites. One way to remember the difference is to associate the letter *g* in stalagmite with "ground"—stalagmites point up from the ground. Stalactites and stalagmites can grow together forming columns.

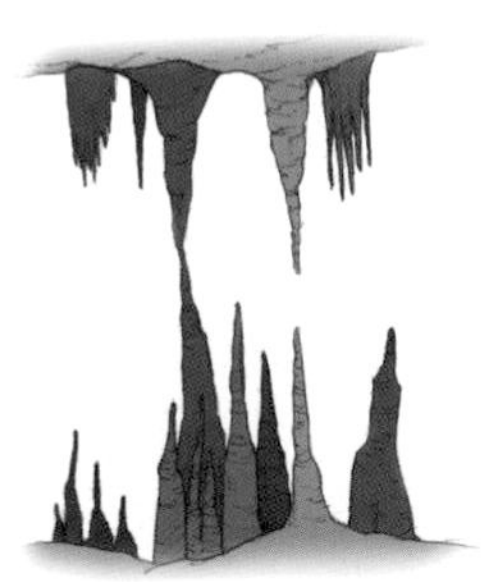

Sinkholes

Sedimentary terrain

Sinkholes are areas where part of the bedrock below the land surface has dissolved and the land surface has collapsed into the space below, sometimes catastrophically.

KEY FACTS

Bowl-shaped depressions in the land surface come in various sizes, some as large as hundreds of feet in diameter.

+ fact: Features of karst landscape

+ fact: Can develop rapidly, especially after a heavy rain

+ fact: Common in Florida, Texas, Alabama, Missouri, Kentucky, Tennessee, and Pennsylvania

Sinkholes are landforms that result from a collapse of the land surface. They occur in regions underlain by bedrock that is easily dissolved in groundwater. Examples are carbonate rocks and evaporite deposits such as salt and gypsum. The movement of groundwater causes the rock to dissolve over time, slowly enough that the ground above may stay intact. Eventually, however, dissolution leaves the bedrock unable to support the surface, and the land collapses into the hole. This can happen on various scales and can be catastrophic if buildings or roads are built on the affected surface.

Desert Pavement

Desert environments

Desert pavement is a hard ground surface made of tightly packed pebbles and cobbles overlying a layer of fine sediment. Desert pavements are found in arid and semiarid environments.

KEY FACTS

A thin layer of packed rock fragments and clasts covers the ground.

+ fact: Many clasts are coated with a dark-colored patina called desert varnish.

+ fact: Exposures range from small patches to areas as large as hundreds of square miles.

+ fact: Newspaper Rock in Utah is a famous example.

Desert pavement is a ground surface common in some windblown deserts. It is formed by a thin layer of tightly packed clasts, pebble to cobble size, and generally of various shapes, sizes, and types. Ventifacts—wind-shaped particles—are common among these clasts. Desert pavements are found worldwide, even in Antarctica. The manner of their formation is controversial. Various ideas proposed to explain the features include deposition of the clasts during catastrophic rain events, uplift of the clasts due to wetting and drying or freezing and thawing processes, or creation of the pavement in place by wind erosion.

Ventifacts

Desert environments

Ventifacts are rocks shaped, polished, and etched by windblown sand. They come in all shapes and sizes, from small, polished, aerodynamic pebbles to large mushroom-shaped boulders.

KEY FACTS

These rocks in desert environments have smoothed, polished surfaces and/or pitted or fluted surfaces created by wind erosion.

+ fact: Look for ventifacts in exposures of desert pavement.

+ fact: The shape can indicate wind direction.

+ fact: Ventifacts have been identified on Mars and in Antarctica.

Ventifacts are formed in arid environments by the abrasive power of sand and smaller particles carried by the wind. They have various forms, depending on original characteristics of the rocks and energy and direction of the wind. Some ventifacts have smooth, wind-blasted faces. Others contain surfaces with networks of grooves and pits. Some ventifacts contain several wind-smoothed faces, suggesting that either the rock moved during wind abrasion or the wind direction changed during the ventifact's formation. Ancient ventifacts preserved in beds of sedimentary rock are indications of a stony desert in the past.

Alluvial Fans

Desert environments

Alluvial fans are cones of sediment similar to deltas, which form where there is an abrupt change in topography from steep uplands to a flat plain.

KEY FACTS

These are fan- or cone-shaped accumulations of gravel, sand, and smaller materials.

+ fact: Made of unconsolidated sediment known as alluvium

+ fact: Identified on Mars

+ fact: Spectacular alluvial fans are found in Death Valley National Park.

Alluvial fans are characteristic of arid and semiarid environments in which mountains or hills abut a wide basin or plain. The high ground is drained along valley systems to the basin, where there is an abrupt change in topography. Here, flows of water and sediment rapidly lose energy and the sediment load is deposited. Repeated deposition builds up cones of sediment at the mouth of drainages. Alluvial fans are similar to deltas but occur above water in arid environments. Alluvial fans from neighboring valleys can converge when built up with enough sediment.

Sand Dunes

Desert environments

Sand dunes are distinctive features of sandy deserts, formed by wind-powered transport and deposition of sand grains. Dune shapes depend on wind direction and sand abundance.

KEY FACTS

Dunes are accumulations of sand grains in piles with straight or curved crests.

+ fact: Some cross beds in sandstones are the mark of ancient sand dunes.

+ fact: Colorado's Great Sand Dunes National Park contains the tallest sand dunes in North America.

+ fact: Sand dunes are features of expansive sandy deserts called ergs.

Arid deserts are characterized by landforms built from wind energy. As wind blows across sand, sand grains are progressively skipped forward by a process known as saltation. Saltating sand grains accumulate until the crest of the pile is unstable and the grains avalanche down the lee side, forming the shape of a dune. Straight-crested dunes called "transverse dunes" form perpendicular to the prevailing wind direction. When there is limited sand supply, individual half-moon–shaped dunes called "barchan dunes" form. When there are two or more prominent wind directions, dunes with different orientations and shapes form.

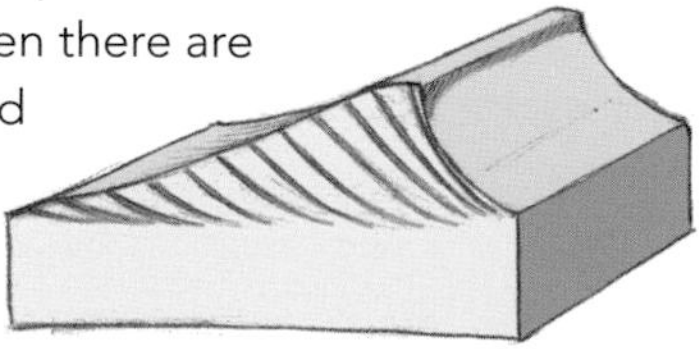

Meandering Rivers

River environments

Rivers that flow in sinuous channels with many bends and loops are called meandering rivers. These rivers tend to occur in low areas with a relatively consistent elevation gradient.

KEY FACTS

Loopy, sinuous channels are signs of meandering rivers.

+ fact: Meandering rivers have wide floodplains.

+ fact: Channels of meandering rivers are deepest on the outside of each bend.

+ fact: The Mississippi River in the southeastern United States is a classic meandering river.

Two primary river patterns account for side-to-side movement of a river course: meandering and braided rivers. Meandering rivers form in areas with a fairly consistent and low topographic gradient. The current occupies a single channel and swings back and forth across bends and loops on each side of its overall flow direction. The river scours the outside of the bends and deposits material on the inside, deepening the loops over time. Eventually, the current can break through adjacent loops of the deepest meanders, causing the channel to shortcut and change course. The cutoff meander becomes an oxbow lake.

Braided Rivers

River environments

Rivers in a network of interconnected channels are called braided rivers or streams. These tend to form with a sudden change in the elevation gradient and/or the sediment load.

KEY FACTS

Rivers with multiple channels converge and diverge in a braided pattern.

+ fact: Found in areas of relatively steep topography

+ fact: Common in regions with glacial outwash

+ fact: Beautiful examples of braided rivers are found in Alaska.

When the elevation gradient of a river changes suddenly, as in a transition from a mountainous region to a broad plain, the river pattern that develops is called braided. In this pattern, the river occupies numerous channels that branch off from one another, converge, and then diverge again irregularly. Braided streams are also characterized by a high sediment load. They leave behind stones and gravels concentrated in high strips that divide channels, and silts and clays deposited in hollows between them. Braided streams are common in outwash plains and alluvial fans at the foot of mountains.

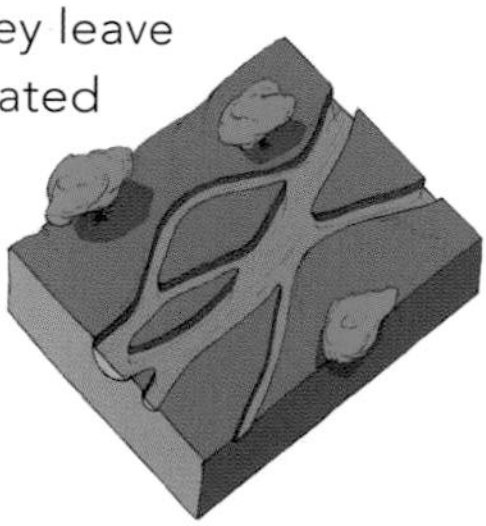

Deltas

River environments

Deltas are broad regions of deposited sediment at river mouths. They form when rivers suddenly lose their gradient and deposit their sediment load at the coast of a sea or lake.

KEY FACTS

Broad accumulations of sediment occur where rivers empty into oceans or lakes.

+ fact: Sediment becomes finer grained along the far edges of the delta.

+ fact: River deltas create important agricultural lands.

+ fact: Some ancient delta sediments are petroleum reservoirs.

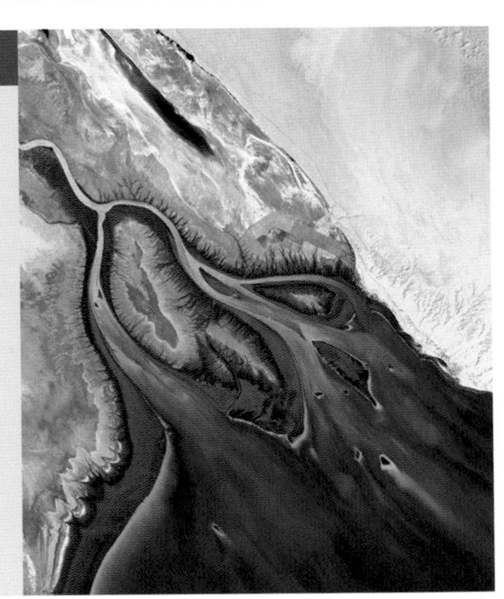

Deltas are broad deposits of fine sediment, mostly silt, at the mouth of a river where it meets a lake or sea. The river loses its gradient and rapidly slows, and this decrease in energy causes the river to drop its sediment load. Deltas are similar to alluvial fans, but occur along coastlines, and the deposited materials are finer because they are farther from the source. Deltas are named after the Nile River Delta, whose shape resembles the triangular capital letter delta in the Greek alphabet. Deltas have different shapes. The Mississippi Delta is called a "bird's foot delta" because of its fingerlike projections of sediment.

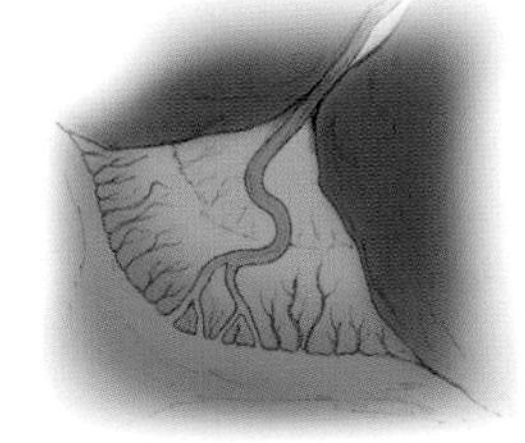

Potholes

River environments

Potholes are smooth, circular depressions in surfaces of exposed bedrock. They form by the abrasive process of sand and gravel whirled around a depression by flowing water.

KEY FACTS

Circular depressions appear in bedrock, some with polished surfaces.

+ fact: Some are called giants' cauldrons or kettles.

+ fact: Many in the U.S. and Canada formed by glacial meltwaters during the last ice age.

+ fact: Some of the world's deepest are found in basalt at Interstate State Park in Minnesota.

Potholes are rounded depressions in rock, with shapes and sizes ranging from small, cylindrical holes to gigantic, cauldron-shaped bowls. Most potholes are formed by river abrasion: the scouring action of stream-carried sediment against exposed bedrock. Sand and gravel, trapped in small fissures and depressions, are whirled by the current, scouring out potholes. This process is greatest during floods, when energy is high enough to transport heavy, coarse sediment with more erosive power. Potholes can indicate previous glacial activity because meltwater from a receding glacier is a powerful scouring agent.

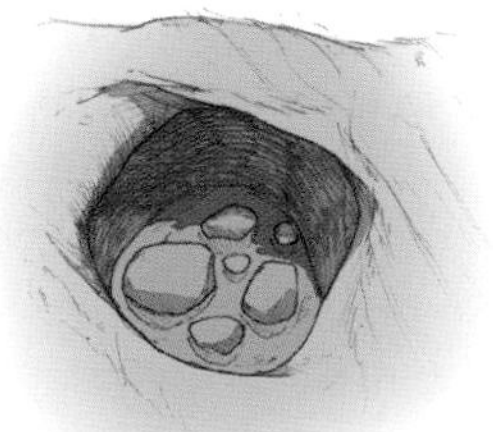

Barrier Islands

Coastal environments

Barrier islands are sandbars or sandspits not connected to a coastline. They are separated from the coast by a sound or a lagoon, and they are high enough for dunes to form.

KEY FACTS

Sandy islands are separated from the continent by a sound, bay, or lagoon.

+ fact: Barrier islands move and change shape with ocean currents and storm events.

+ fact: Vegetation helps stabilize barrier islands.

+ fact: The Outer Banks of North Carolina are barrier islands.

Barrier islands are linear deposits of sand offshore, separated from the mainland by a sound, bay, or lagoon. They typically have a sandy beach on the seaward side, then an area of dunes, and then a marsh leading to a shallow lagoon on the inland side. Barrier islands are somewhat stabilized by dune vegetation, but they can change dramatically from season to season and storm to storm. Barrier islands "migrate" in two ways: Alongshore ocean currents move sediment from one tip of the island to the other, or storms remove beach sand from the ocean side and deposit it on the marsh side as overwash.

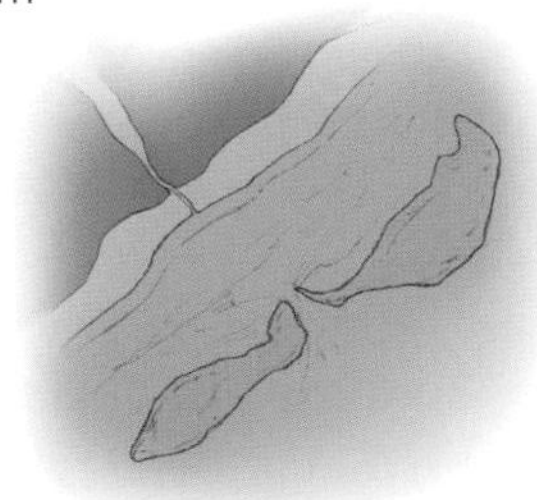

Shoals/Sandspits/Sandbars

Coastal environments

Shoals, sandspits, and sandbars are coastal landforms that develop where streams and alongshore currents create linear deposits of sand and mud.

KEY FACTS

These landforms consist of linear deposits of sand and mud.

+ fact: Shoals create local areas of shallow water and are navigation hazards for boats.

+ fact: A nearly continuous chain of sandspits and barrier islands extends from Long Island, New York, to Florida.

+ fact: Cape Cod, Massachusetts, is a sandspit.

Shoals are elongate deposits of sand and mud that extend into water in coastal environments. Shoals that are attached to a headland but that extend into the water past a headland bend are called spits or sandspits. These form via alongshore drift. When waves meet the beach at an oblique angle but recede perpendicularly, the change in current direction transports sand down the shore. When the angle of a headland changes, alongshore drift transports sand past the bend, extending the sand deposit into the sea until the current loses energy. Sandbars are isolated shoals detached from headlands and surrounded by water.

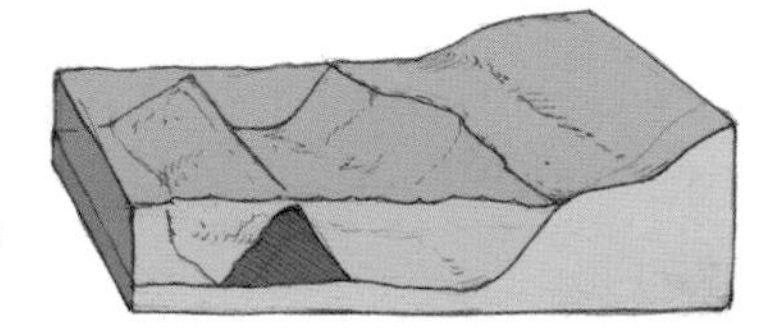

Marine Terraces or Platforms

Coastal environments

Marine terraces or platforms are bench-like coastal landforms that usually indicate a lowering of sea level or a rise of the land. These terraces are cut by wave erosion before they are exposed.

KEY FACTS

Broad "benches" are found along coasts.

+ fact: Some terraces along the California coast are uplifted by movement along the San Andreas Fault.

+ fact: Dramatic stacked terraces are exposed in the Palos Verdes Hills in Los Angeles County, California.

+ fact: Common on the Pacific coast of North America

Marine terraces are horizontal platforms separated by steep cliff faces on exposed coastlines. Ocean waves are one of nature's most powerful forces. Wave energy becomes concentrated on exposed headlands, cutting into the rock like a horizontal saw, causing overlying rock to collapse and retreat landward. This process cuts a nearly horizontal platform adjacent to a steep scarp. If the land rises with respect to sea level, this platform is exposed above the water. Periodic uplift (or sea level decline) will expose marine terraces in a steplike pattern, as along the coast of California.

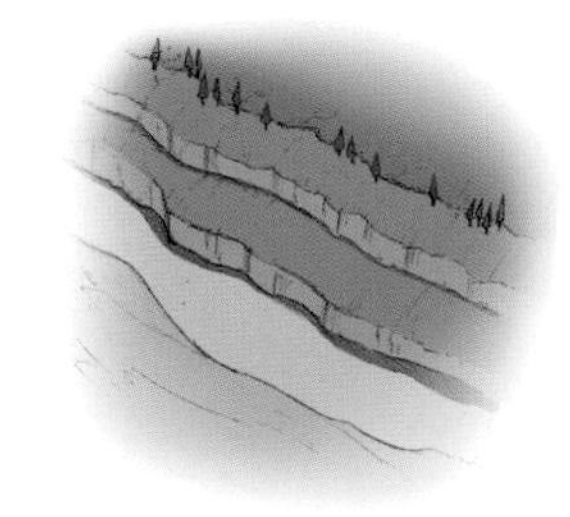

Sea Stacks

Coastal environments

A sea stack is a remnant of a coastal headland, carved by the powerful erosive force of ocean waves. Sea stacks generally form where resistant bedrock is exposed along a high-energy coast.

KEY FACTS

Isolated rock towers rise out of the ocean along a coastline.

+ fact: Can form from many different types of rock, including limestone, sandstone, and volcanic tuff

+ fact: Sea stacks are important bird habitats.

+ fact: Common along the West Coast; famous along Harris Beach, Oregon

Sea stacks are isolated towers or monoliths that rise steeply out of the sea. They are generally clustered along steep coastlines marked by resistant bedrock. Sea stacks often form when waves attack jointed or faulted rocks, preferentially eroding along these weak planes, leaving pillars or towers surrounded by seawater. Eventually, wave attack will continue to erode the stacks, and they will topple into the ocean and become pulverized to smaller and smaller particles. Sea stacks can also form in limestone bedrock by the combined effects of wave attack and limestone dissolution.

Cirques

Glaciated terrain

Cirques are bowl-shaped scoops on mountainsides formed by glaciers. Glaciers carve cirques backward into the mountainside by breaking down and removing rock beneath the ice.

KEY FACTS

Bowl-shaped depressions appear in the sides of mountains.

+ fact: Cirques are named from the French word for "amphitheater."

+ fact: Those found in nonglaciated regions are important evidence for previous ice ages.

+ fact: Cirques filled with lakes are called tarns.

Mountain glaciers form along the flanks of high peaks and are relatively small compared with continental glaciers or ice sheets. The bowl-shaped depressions at the head of a mountain glacier are called cirques. Cirques are easily recognizable by their distinctive rounded, hollowed-out form, like the path of a gigantic ice cream scoop on the side of a mountain. Cirques form as the glacial ice and meltwater below the ice gouge out the rock behind the glacier. When two cirques on different sides of a mountain erode deeply enough to intersect, they form a sharp ridge between them called an arête.

Glacial Erratics

Glaciated terrain

Glacial erratics are rocks that have been deposited by a glacier, usually after being transported great distances from their bedrock source.

KEY FACTS

Erratics are out of place in their current environment due to transport by a glacier.

+ fact: Can range in size; boulders are well known.

+ fact: The largest known erratic in the U.S. and Canada is New Hampshire's Madison Boulder, weighing almost 6,000 tons.

+ fact: Some boulders are perched precariously on top of smaller stones.

Glacial erratics are clasts of rock, including large boulders, transported and deposited by glaciers. They are called erratics because they are incongruous with the surrounding bedrock, having been moved sometimes great distances by a glacier. Scientists use erratics to map the extent of the continental ice sheet that covered much of Canada and the U.S. during the last ice age, which reached its maximum extent approximately 21,000 years ago. Glacial erratics originate when glaciers pluck rocks as they scour over topographic highlands. When the glacier melts, it drops its sediment load into an unconsolidated pile.

Kettle Ponds

Glaciated terrain

Kettle holes or ponds are depressions formed by melting chunks of glacial ice that are no longer connected to the glacier. After the glacier has receded, these holes often fill with water.

KEY FACTS

Kettle ponds are depressions formed in glacial outwash plains that have filled in with water.

+ fact: The ponds are recharged by groundwater and/or rainwater.

+ fact: Thoreau's Walden Pond in Massachusetts is a famous kettle pond.

+ fact: The ponds of Cape Cod, Massachusetts, are kettle ponds.

Kettle ponds are important markers of glaciated terrain. These variably sized ponds are found in regions that were previously glaciated. They form when chunks of ice calve off the front of a receding glacier and are buried by glacial sediment called till. When the chunks of ice melt away, the till collapses into the space, creating a hole. Over time, these holes fill with water and become ponds. Kettle ponds are common landforms across the northern region of North America because of the continental ice sheet that reached its maximum extent approximately 21,000 years ago during the last glacial period.

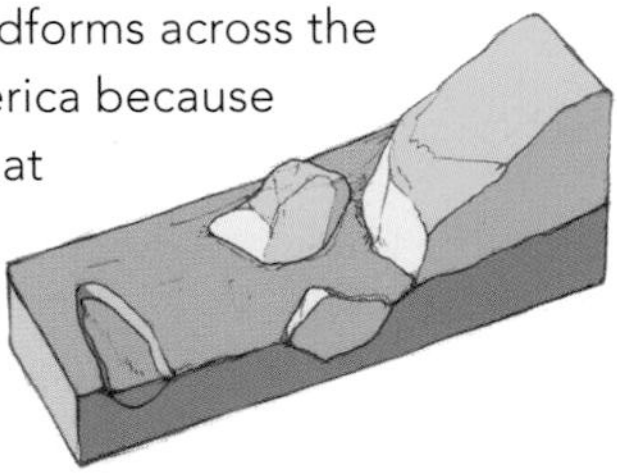

U-shaped Valleys

Glaciated terrain

Valleys carved by glaciers develop a characteristic U-shaped cross section. U-shaped valleys are key indicators of glacial erosion, whereas V-shaped valleys are carved by rivers.

KEY FACTS

Broad troughs in the landscape have U-shaped cross sections.

+ fact: U-shaped valleys filled with seawater are called fjords.

+ fact: U-shaped valleys are important indicators of past glacial periods.

+ fact: Fantastic examples can be seen in Glacier National Park in Montana.

Distinctive markers of glaciated terrain are steep-walled valleys with broad floors that resemble the shape of a U in cross section. U-shaped valleys are formed by the downhill flow of glaciers. The ice carves out steep walls and a relatively flat base. These valleys are also often straighter in course than river-cut valleys. Some U-shaped valleys can be notched by V-shaped river channels from the flow of glacial meltwater. The floors of tributary glaciers can be significantly higher than the main valley floor, leading to waterfalls and hanging valleys after the retreat of the glaciers.

Natural Regions of North America

The continental United States and Canada can be divided into nine physiographic regions.

Appalachian Highlands

The oldest mountain chain in North America, the heavily eroded Appalachians extend for about 2,000 miles (3,200 km) from Alabama to Newfoundland, with the highest peak being Mount Mitchell, in North Carolina, at 6,684 feet (2,040 m). This mountain range started forming 480 million years ago, when continental collisions caused volcanic activity and mountain building. Ecologically, this region hosts eastern temperate forests with a great variety of coniferous and deciduous trees and wildflowers, some of them high-elevation specialists.

Coastal Plain

This gradually rising flatland spans some 4,000 miles (6,400 km) in total and covers several distantly related regions along the Gulf of Mexico, the southern Atlantic coast, and the northernmost coasts of the Arctic Ocean. In the past, oceans covered these plains, depositing sediment layers over millions of years, until falling sea levels exposed them. The types of vegetation in these widely separated regions range from subtropical trees and flowers in the southernmost coastal plain to marsh plants along the mid-Atlantic to the largely treeless tundra on the northern coast of Alaska.

Interior Plains

Ranging from the lowlands of the Saint Lawrence River Valley in the east to the mile-high Great Plains in the west, the vast Interior Plains of North America provide fertile soils, especially for productive prairie farms. A shallow sea covered much of this region as recently as 75 million years ago, and sediments from rivers draining the Appalachians and western mountains were deposited in layers throughout the sea. The Great Plains were once covered by vast and diverse expanses of natural grasses, sagebrush, and a varied suite of wildflowers. Much of this ecosystem has vanished, the land brought into use by modern agriculture and extensive grazing.

Interior Highlands

The Ozark Plateau and Ouachita Mountains form the Interior Highlands, which are centered on Arkansas and southern Missouri, with mountains reaching more than 2,600 feet (800 m) high. These ancient eroded highlands were connected to the Appalachians until tectonic activity separated them some 200 million years ago. Ecologically, this relatively small area straddles the southern Interior Plains and the eastern Coastal Plain, with trees and wildflowers representing both regions.

Rocky Mountains

The highest mountain system in North America, the Rockies dominate the landscape for some 3,000 miles (4,800 km), from New Mexico to Alaska, with more than 50 peaks surpassing 14,000 feet (4,300 m). Tectonic activity uplifted the Rockies about 50 to 100 million years ago, making them much younger and less eroded than the Appalachians. The ecological hallmarks of the Rockies are its coniferous forests of pines, firs, and spruces, adapted to high elevations, with wildflower species similarly adapted to elevations and temperatures.

Intermontane Basins and Plateaus

This region is called intermontane because it is situated between the Pacific and the Rocky Mountain systems. Pacific mountains block most moisture-bearing clouds coming from the Pacific Ocean, giving desert climates to places like the Colorado Plateau, 5,000 to 7,000 feet (1,500 to 2,100 m) high, and the Great Basin. In Canada and Alaska, the immense Yukon River Valley and the Yukon-Tanana Uplands are part of this region. The deserts feature cacti and a host of other specialist plants of the arid West. Cottonwoods, ashes, and willows line rivers that run intermittently through the dry plains.

Pacific Mountain System

From Alaska to California, mountains and volcanoes tower over the West Coast in an almost unbroken chain. These mountain ranges are geologically young and seismically active, with uplift starting some five million years ago. The highest mountain in North America, Alaska's Mount McKinley (20,320 ft/6,200 m), is still growing at about a millimeter a year—about the thickness of a fingernail. Distantly separated from the Rockies, this mountain range supports its own distinct varieties of trees and wildflowers adapted to higher elevations.

Canadian Shield

The geologic core of North America is the Canadian Shield, which contains the continent's oldest rocks. Landforms are relatively flat, having been eroded and scoured by glaciers over millions of years. The exposed bedrock ranges in age from 570 million to more than 3 billion years old. This is a vast and extensively diverse region of climatic extremes and varied vegetation, from dense boreal forests in the south to frigid tundra in the north, populated by stunted trees, small shrubs, lichens, and ground-clinging herbs.

Arctic Lands

Highlands known as the Innuitian Mountains cover most islands. The icy climate is too harsh for most animals and vegetation, and much of the ground is permanently frozen. Nevertheless, low-growing shrubs, small tundra plants, and lichens manage to survive.

Further Resources

WEBSITES

Association of American State Geologists
www.stategeologists.org

Dinosaur Society's Dinosaur News
www.dinosaursociety.com/news/category/dinosaur-news

Geological Society of America
www.geosociety.org

Mindat.org
www.mindat.org

Paleontology Portal
www.paleoportal.org

U.S. Geological Survey
www.usgs.gov

U.S. Geological Survey Volcano Hazards Program
www.volcanoes.usgs.gov

University of California Museum of Paleontology's Online Exhibits
www.ucmp.berkeley.edu/exhibits/index.php

Volcano World
http://volcano.oregonstate.edu

About the Author

SARAH GARLICK is a writer and educator specializing in earth and environmental science. She holds degrees in geology from Brown University and the University of Wyoming, and she is the author of the award-winning book *Flakes, Jugs, and Splitters: A Rock Climber's Guide to Geology* (2009). She lives with her family in the mountains of New Hampshire.

Acknowledgments

The following geologists provided helpful reviews of this project at various stages: Mark Clementz, Ron Frost, Peter Isaacson, Ranie Lynds, Joshua Schwartz, Arthur Snoke, and Victor Zabielski. I'd like to thank each of them for their time and interest. I'd also like to extend my thanks and acknowledgment to the editorial team at National Geographic Books, especially Barbara Payne and Paul Hess, and to Jared Travnicek for providing illustrations. Finally, my gratitude goes to my husband, Jim Surette, for his enduring love and support.

About the Artist

JARED TRAVNICEK is a scientific and medical illustrator. He received his M.A. in Biological and Medical Illustration from the Johns Hopkins University School of Medicine in Baltimore, Maryland. Travnicek is a Certified Medical Illustrator and a professional member of the Association of Medical Illustrators.

Illustrations Credits

Abbreviations for terms appearing below:
Getty Images (GI), Science Source (SS), Smithsonian Institution (SI), Visuals Unlimited, Inc. (VU)

All artwork appearing in this book was created by Jared Travnicek, Jenny Wang, and Fernando G. Baptista.

FRONT COVER
UP: Mark Thiessen, NGS; LO (left to right): Charles Schug/GI; Herman Eisenbeiss/GI; Pasieka/SS; Mark Schneider/GI.

SPINE
Gary Retherford/GI.

BACK COVER
(UPLE), Biophoto Associates/GI; (UPRT), Thomas P. Shearer/GI; (CTR), Shoshannah White/GI; LO (left to right): Gavin Kingcome/SPL/GI; Darrell Gulin/GI; Vaughan Fleming/SS; Dirk Wiersma/SS.

1, Mark Schneider/VU; 2-3, Don White/SuperStock/Corbis; 4, Art Wolfe/www.artwolfe.com; 5, Boris Sosnovyy/Shutterstock; 7, Pasieka/SS; 10, Jason Patrick Ross/Shutterstock; 12 (UP), Juraj Kovac/Shutterstock; 12 (LO), Boris Sosnovyy/Shutterstock; 13 (UP), Scientifica/VU; 13 (LO), DEA/A.RIZZI/GI; 14 (UP), Joel Arem/SS; 14 (LO), Mark A. Schneider/SS; 15 (UP), Scott Camazine/SS; 15 (LO), 2013 National Museum of Natural History, SI; 16 (UP), Scientifica/VU; 16 (LO), The Trustees of the British Museum/Art Resource, NY; 17 (UP), Mark Schneider/VU; 17 (LO), David McLain/Aurora/GI; 18 (UP), Mark A. Schneider/SS; 18 (LO), Biophoto Associates/SS; 19 (UP), Andrew J. Martinez/SS; 19 (LO), Phil Degginger; 20 (UP), Marli Miller/VU; 20 (LO), Mark Steinmetz; 21 (UP), Biophoto Associates/SS; 21 (LO), 2013 National Museum of Natural History, SI; 22 (UP), Ted Kinsman/SS; 22 (LO), Biophoto Associates/SS; 23 (UP), Marli Miller/VU; 23 (LO), Joel Arem/SS; 24 (UP), Charles D. Winters/SS; 24 (LO), The Trustees of the British Museum/Art Resource, NY; 25 (UP), 2013 National Museum of Natural History, SI; 25 (LO), Joel Arem/SS; 26 (UP), John Cancalosi/GI; 26 (LO), Richard Leeney/GI; 27 (UP), RF Company/Alamy; 27 (LO), Biophoto Associates/SS; 28 (UP), Gary Cook/VU; 28 (LO), James H. Robinson/SS; 29 (UP), DEA/Photo 1/VU; 29 (LO), Harry Taylor/Dorling Kindersley/GI; 30 (UP), Joel Arem/SS; 30 (LO), Boykung/Shutterstock; 31 (UP), Roberto de Gugliemo/SS; 31 (LO), Image copyright © The Metropolitan Museum of Art. Image source: Art Resource, NY; 32 (UP), Biophoto Associates/SS; 32 (LO), Marli Miller/VU; 33 (UP), Dirk Wiersma/SS; 33 (LO), Tyler Boyes/Shutterstock; 34 (UP), Scientifica/VU; 34 (LO), Siim Sepp/Alamy; 35 (UP), Manamana/Shutterstock; 35 (LO), Biophoto Associates/SS; 36 (UP), 2013 National Museum of Natural History, SI; 36 (LO), Dirk Wiersma/SS; 37 (UP), Joel Arem/SS; 37 (LO), Scientifica/VU; 38 (UP), Ted Kinsman/SS; 38 (LO), Image copyright © The Metropolitan Museum of Art. Image source: Art Resource, NY; 39 (UP), Scientifica/VU; 39 (LO), WitthayaP/Shutterstock; 40 (UP), 2013 National Museum of Natural History, SI; 40 (LO), bpk, Berlin/Aegyptisches Museum, Staatliche Museen, Berlin, Germany/Margarete Buesing/Art Resource, NY; 41 (UP), Scientifica/VU; 41 (LO), Mark A. Schneider/SS; 42 (UP), Nadezda Boltaca/Shutterstock; 42 (LO), 2013 National Museum of Natural History, SI; 43 (UP), 2013 National Museum of Natural History, SI; 43 (LO), Scala/Art Resource, NY; 44 (UP), 2013 National Museum of Natural History, SI; 44 (LO), Tyler Boyes/iStockphoto; 45 (UP), Scientifica/VU; 45 (LO), E. R. Degginger/SS; 46 (UP), carlosdelacalle/Shutterstock; 46 (LO), Mark Schneider/VU; 47 (UP), 2013 National Museum of Natural History, SI; 47 (LO), GIPhotoStock/SS; 48 (UP), 2013 National Museum of Natural History, SI; 48 (LO), Mark Schneider/VU; 49 (UP), 2013 National Museum of Natural History, SI; 49 (LO), The Trustees of the British Museum/Art Resource, NY; 50 (UP), 2013 National Museum of Natural History, SI; 50 (LO), Album/Art Resource, NY; 51 (UP), 2013 National Museum of

Natural History, SI; 51 (LO), Roman Beniaminson/Art Resource, NY; 52 (UP), 2013 National Museum of Natural History, SI; 52 (LO), AnatolyM/Shutterstock; 53 (UP), Vitaly Raduntsev/Shutterstock; 53 (LO), Scientifica/VU; 54 (UP), 2013 National Museum of Natural History, SI; 54 (LO), Martin Novak/Shutterstock; 55 (UP), George Whitely/SS; 55 (LO), 2013 National Museum of Natural History, SI; 56 (UP & LO), Scientifica/VU; 57 (UP), Doug Martin/SS; 57 (LO), Dr. Marli Miller/VU; 58 (UP), Depositphotos.com/Borislav Marinic; 58 (LO), Dr. Marli Miller/VU; 59 (UP), Scientifica/VU; 59 (LO), Marli Miller/VU; 60 (UP), Marli Miller/VU; 60 (LO), Mark Schneider/VU; 61 (UP), Dan Suzio/SS; 61 (LO), Ron Schott; 62 (UP), Spring Images/Alamy; 62 (LO), Dirk Wiersma/SS; 63 (UP), Marli Miller/VU; 63 (LO), Scientifica/VU; 64 (UP), Duncan Shaw/SS; 64 (LO), Doug Martin/SS; 65 (UP), Peter von Bucher/Shutterstock; 65 (LO), Ted Kinsman/SS; 66 (UP), Ron Schott; 66 (LO), Michael Szoenyi/SS; 67 (UP), Alan Majchrowicz; 67 (LO), E. R. Degginger/SS; 68 (UP), William H. Mullins/SS; 68 (LO), Wally Eberhart/VU; 69 (UP), Jerry McCormick-Ray/SS; 69 (LO), kavring/Shutterstock; 70 (UP), Gerald & Buff Corsi/VU; 70 (LO), Adrienne Hart-Davis/SS; 71 (UP), David Hosking/SS; 71 (LO), Joyce Photographics/SS; 72 (UP), Images & Volcans/SS; 72 (LO), DEA/R. APPIANI/GI; 73 (UP), Dr. Marli Miller/VU; 73 (LO), Joyce Photographics/SS; 74 (UP), John Buitenkant/SS; 74 (LO), Mark A. Schneider/SS; 75 (UP), Inga Spence/VU; 75 (LO), Tyler Boyes/Shutterstock; 76 (UP), Francois Gohier/SS; 76 (LO), michal812/Shutterstock; 77 (UP), Richard J. Green/SS; 77 (LO), E. R. Degginger/SS; 78 (UP), Steve McCutcheon/VU; 78 (LO), Ted Kinsman/SS; 79 (UP), Dr. Marli Miller/VU; 79 (LO), Biophoto Associates/SS; 80 (UP), Christian Grzimek/SS; 80 (LO), Scientifica/VU; 81 (UP), Hubertus Kanus/SS; 81 (LO), Scientifica/VU; 82 (UP), Len Rue Jr./SS; 82 (LO), Albert Copley/VU; 83 (UP), Gianni Tortoli/SS; 83 (LO), Trevor Clifford Photography/SS; 84 (UP), Ken M. Johns/SS; 84 (LO), Joyce Photographics/SS; 85 (UP), Ken M. Johns/SS; 85 (LO), Joyce Photographics/SS; 86 (UP), Michael P. Gadomski/SS; 86 (LO), Scientifica/VU; 87 (UP), Ellen Thane/SS; 87 (LO), Biophoto Associates/SS; 88 (UP), Dr. Marli Miller/VU; 88 (LO), Tyler Boyes/Shutterstock; 89 (UP), John Arnaldi/VU; 89 (LO), Marli Miller/VU; 90 (UP), Arthur W. Snoke; 90 (LO), 2013 National Museum of Natural History, SI; 91 (UP & LO), Marli Miller/VU; 92 (UP), Marli Miller/VU; 92 (LO), Joel Arem/SS; 93 (UP & LO), Scientifica/VU; 94 (UP), Albert Copley/VU; 94 (LO), Scientifica/VU; 95 (UP), Marli Miller/VU; 95 (LO), Andrew Alden; 96 (UP), William D. Bachman/SS; 96 (LO), Aaron Haupt/SS; 97 (UP), John Shaw/SS; 97 (LO), © RMN-Grand Palais/Art Resource, NY; 98 (UP), Walt Anderson/VU; 98 (LO), Image copyright © The Metropolitan Museum of Art. Image source: Art Resource, NY; 99 (UP), Jack Ballard/VU; 99 (LO), Albert Copley/VU; 100, Mark A. Schneider/SS; 101, Barbara Strnadova/SS; 102, Colin Keates/GI; 103, Mark A. Schneider/SS; 104, Mark A. Schneider/SS; 105 (UP), Scott Camazine/SS; 105 (LO), Trilobite illustration (c) Emily S. Damstra; 106, Mark A. Schneider/SS; 107, Mark A. Schneider/SS; 108, Mark A. Schneider/SS; 109, Holbox/Shutterstock; 110, Mark A. Schneider/SS; 111, Scott Camazine/SS; 112, Dirk Wiersma/SS; 113, Geoff Kidd/SS; 114, James L. Amos/SS; 115, Volker Steger/SS; 116, Francois Gohier/SS; 117, Dr. John D. Cunningham/VU; 118, John Cancalosi/Okapia/SS; 119 (UP), Victor Habbick Visions/SS; 119 (LO), Francois Gohier/SS; 120, Marli Miller/VU; 121, Marli Miller/VU; 122, Marli Miller/VU; 123, Marli Miller/VU; 124, Michael Szoenyi/SS; 125, Marli Miller/VU; 126, Doug Sokell/VU; 127, Ted Kinsman/SS; 128, Dennis Flaherty/SS; 129, Bruce M. Herman/SS; 130, Dr. Ken Wagner/VU; 131, Marli Miller/VU; 132, mikenorton/Shutterstock; 133, EastVillage Images/Shutterstock; 134, Pierre Leclerc/Shutterstock; 135, Jim Edds/SS; 136, Ken M. Johns/SS; 137, Marli Miller/VU; 138, Marli Miller/VU; 139, Walt Anderson/VU; 140, Bryan Lowry/SeaPics.com; 141, Stephen & Donna O'Meara/SS; 142, Explorer/SS; 143, Stephen & Donna O'Meara/SS; 144, Georg Gerster/SS; 145, Georg Gerster/SS; 146, Francois Gohier/SS; 147, Brenda Tharp/SS; 148, Inga Spence/VU; 149, Marli Miller/VU; 150, Robert and Jean Pollock/SS; 151, Marli Miller/VU; 152, Craig K. Lorenz/SS; 153, William D. Bachman/SS; 154, Douglas Knight/Shutterstock; 155, Jim W. Grace/SS; 156, ANT Photo Library/SS; 157, Marli Miller/VU; 158, Marli Miller/VU; 159, Tim Pleasant/Shutterstock; 160, Michael Male/SS; 161, Mark Newman/SS; 162, Planet Observer/SS; 163, Michael P. Gadomski/SS; 164, Marli Miller/VU; 165, William D. Bachman/SS; 166, G. R. 'Dick' Roberts/NSIL/VU; 167, G. R. 'Dick' Roberts/NSIL/VU; 168, Ned Therrien/VU; 169, Andrew J. Martinez/SS; 170, Thomas & Pat Leeson/SS; 171, Bill Kamin/VU.

Index

Boldface indicates mineral, rock, fossil, or landform profile.

A

A'a lava **140**
Agate **14**
Alabaster 43
Alluvial fans **158**
Alluvium 158
Amber 114
Amethyst 6, 12
Ammonites **109**
Amphibians **115**
Amphibole **33**
Amphibolite **90**
Andalusite **25**
Andesite **73**
Anhydrite **43**
Anorthosite **65**
Anticlines 124
Antigorite 38
Aplite **59**
Aquamarine 30
Aquifers, karst 153
Archaeopteryx 117
Arches, natural **132**
Arenites, quartz **77**
Arêtes 168
Arkose **79**
Arthropods 105
Asbestos 38
Augen **122**
Augite 32
Aventurine 99
Azurite 42

B

Barrier islands **164**
Basalt **75**
Bedding **148**; *see also* Cross bedding
Beryl **30**
Biotite **37**
Bird fossils **117**
Bivalves **107**
Bloodstone 13
Blueschist **95**
Bogs 86
Boudinage **123**
Brachiopods **106**
Braided rivers **161**
Breccia **85**
Bridges, natural 132
Bryozoans **103**

C

Calcite **40**
Calderas **145**
Carbon, pure 52
Carbonate minerals
 azurite 42
 calcite **40**
 dolomite **41**
 malachite **42**
Carnelian 13
Cavernous weathering. *see* Tafoni
Chalcedony **13**
Chalcopyrite **45**
Chalk **83**
Chert **17**
Chevrons 124
Chiastolite 25
Chrysotile 38
Cinder cones **144**
Cirques **168**
Citrine 12
Coastal environments
 barrier islands **164**
 marine terraces or platforms **166**
 sea stacks **167**
 shoals/sandspits/sandbars **165**
Columnar jointing **147**
Concrete 74
Conglomerate **84**
Copper **51**
Coprolites 119
Corals **104**
Corundum **55**
Creep **135**
Crenulations 124
Crinoids **111**
Cross bedding **150**
Cross lamination. *see* Ripple marks

D

Dacite **72**
Deformed rocks
 augen **122**
 boudinage **123**
 faults **125**
 folds **124**
 foliation **121**
 joints **126**
 lineation **120**
 veins **127**
Deltas **162**
Desert environments
 alluvial fans **158**
 desert pavement **156**
 sand dunes **159**
 ventifacts **157**
Desert pavement **156**
Desert roses 43
Desert varnish **129**
Diabase **74**
Diamond **52**
Dikes **136**
Dinosaurs **116**
Diorite **63**
Disconformities 149
Dolomite (mineral) **41**
Dolomite (rock) **81**
Dolostone **81**
Dome, resurgent 145
Dunes **159**
Dung, fossilized 119
Dunite **67**

E

Echinoderms **110**
Emerald 6, 30
Epidote **29**
Ergs 159
Exfoliation joints **131**

F

Fairy chimneys 134

Faults **125**
Fayalite 23
Feldspars **18**
  labradorite **21**
  plagioclase **20**
  potassium feldspars **19**
Fins, stone **133**
Fish fossils **112**; *see also* Shark teeth
Fissility 86
Fjords 171
Flagstone 76
Flint **17**
Fluorescence 48
Fluorite **48**
Folds **124**
Foliation **121**
Fool's gold. *see* Pyrite
Forsterite 23
Fuchsite 99

G

Gabbro **64**
Galena **44**
Garnet **24**
Garnet schist **93**
Gastropods **108**
Giants' cauldrons 163
Glacial erratics **169**
Glaciated terrain
  cirques **168**
  glacial erratics **169**
  kettle ponds **170**
  U-shaped valleys **171**
Glass, volcanic
  obsidian **70**
  pumice **69**
Glaucophane 95
Gneiss **88**
Gold **50**
Granite 7, **56**
  porphyritic **57**
  rapakivi **58**
Granodiorite **61**
Graptolites **102**
Gravel beds 84
Graywacke **78**
Greenschist **94**
Gypsum **43**

H

Halide minerals
  fluorite **48**
  halite **47**
Halite **47**
Hematite **54**
Honeycombing. *see* Tafoni
Hoodoos **134**
Hornblende **34**

I

Igneous rocks
  andesite **73**
  anorthosite **65**
  aplite **59**
  basalt **75**
  dacite **72**
  diabase **74**
  diorite **63**
  dunite **67**
  gabbro **64**
  granite 7, **56**
  granodiorite **61**
  obsidian **70**
  pegmatite **60**
  peridotite **66**
  porphyritic granite **57**
  pumice **69**
  rapakivi granite **58**
  rhyolite **68**
  syenite **62**
  tuff **71**
Igneous terrain
  a'a lava **140**
  caldera **145**
  cinder cones **144**
  columnar jointing **147**
  dikes **136**
  inclusions **138**
  lava tubes **143**
  pahoehoe lava **141**
  pillow lava **142**
  sills **137**
  volcanic bombs **139**
  volcanic necks or plugs **146**
Inclusions **138**
Insects **114**
Iron roses 54

J

Jasper **16**
Joints **126**
  columnar **147**
  sheeting or exfoliation **131**

K

K-feldspar **19**
Kaolinite **39**
Karst **153**
Kettle ponds **170**
Kyanite **26**

L

L-tectonites 120
Labradorite **21**
Lampshells. *see* Brachiopods
Landslides **135**
Lava
  a'a **140**
  pahoehoe **141**
  pillow **142**
Lava tubes **143**
Lead 44
Limestone **80**
Lineation **120**
Lizardite 38

M

Magnetism 53
Magnetite **53**
Malachite **42**
Mammal fossils **118**
Marble **97**
Marine terraces or platforms **166**
Meandering rivers **160**
Metamorphic rocks
  amphibolite **90**

blueschist **95**
garnet schist **93**
gneiss **88**
greenschist **94**
marble **97**
migmatite **89**
mylonite **91**
quartzite **99**
schist **92**
serpentinite **98**
slate **96**
Mica
biotite **37**
muscovite **36**
Microcline 19
Migmatite **89**
Moonstone 19
Morganite 30
Mudcracks **152**
Mudrocks 86
Mudslides 84
Mudstone **87**
Muscovite **36**
Mushroom rocks **134**
Mylonite **91**

N

Native elements
copper **51**
diamond **52**
gold **50**
silver **49**
Neck, volcanic **146**
Needles, stone 133
Nepheline **22**
Nonconformities 149
Normal faults 125

O

Obsidian **70**
Olivine **23**
Onion-skin weathering. *see* Spheroidal weathering
Onyx 13
Opal **15**
Orthoclase 19
Oxbow lakes 160
Oxide minerals
corundum **55**
hematite **54**
magnetite **53**

P

Pahoehoe lava **141**
Pedestal stones **134**
Pegmatite **60**
Peridot 23
Peridotite **66**
Permineralization 101
Petrified wood **101**
Petroglyphs 129
Phyllite 92
Pillow lava **142**
Plagioclase **20**
Plant fossils **100**; *see also* Petrified wood
Plug, volcanic **146**
Plutonic rock
anorthosite **65**
gabbro **64**
Porphyritic granite **57**
Potassium feldspars **19**
Potholes **163**
Pseudofossils 119
Pumice **69**
Pyrite **46**
Pyroxene **32**

Q

Quartz **12**
Quartz arenite **77**
Quartzite **99**

R

Rapakivi granite **58**
Recumbent folds 124
Red ochre 54
Reptiles **115**; *see also* Dinosaurs
Reverse faults 125
Rhyolite **68**
Ripple marks **151**
River environments
braided rivers **161**
deltas **162**
meandering rivers **160**
potholes **163**
V-shaped valleys 171
Rock salt. *see* Halite
Rubies 55

S

Salt, table. *see* Halite
Salt crystallization 128
Saltation 159
Sand dunes **159**
Sandbars **165**
Sandspits **165**
Sandstone **76**
arkose **79**
graywacke **78**
quartz arenite **77**
Sanidine 19
Sapphires 55
Schist **92**
Schorl 31
Sea stacks **167**
Sedimentary rocks
arkose **79**
breccia **85**
chalk **83**
conglomerate **84**
dolomite/dolostone **81**
graywacke **78**
limestone **80**
mudstone **87**
quartz arenite **77**
sandstone **76**
shale **86**
travertine **82**
Sedimentary terrain
bedding **148**
cross bedding **150**
karst **153**
mudcracks **152**
ripple marks **151**
sinkholes **155**
stalactites and stalagmites **154**
unconformity **149**
Selenite 43
Serpentine **38**
Serpentinite **98**

Shale **86**
Shark teeth **113**
Sheeting joints **131**
Shoals **165**
Silicate minerals
- agate **14**
- amphibole **33**
- andalusite **25**
- beryl **30**
- biotite **37**
- chalcedony **13**
- chert/flint **17**
- epidote **29**
- feldspars **18**
- garnet **24**
- hornblende **34**
- jasper **16**
- kaolinite **39**
- kyanite **26**
- labradorite **21**
- muscovite **36**
- nepheline **22**
- olivine **23**
- opal **15**
- plagioclase **20**
- potassium feldspars **19**
- pyroxene **32**
- quartz **12**
- serpentine **38**
- sillimanite **27**
- staurolite **28**
- talc **35**
- tourmaline **31**

Sillimanite **27**
Sills **137**
Silver **49**
Sinkholes **155**
Slate **96**
Slow creep 135
Slumps **135**
Soapstone 35
Sodalite 62
Spheroidal weathering **130**
Stalactites and stalagmites **154**
Staurolite **28**
Stratovolcanoes 73
Strike-slip faults 125
Sulfate minerals
- anhydrite **43**
- gypsum **43**

Sulfide minerals
- chalcopyrite **45**
- galena **44**
- pyrite **46**

Syenite **62**
Synclines 124

T

Tafoni **128**
Talc **35**
Tarns 168
Tephra. *see* Tuff
Tetrapods 115
Thrust faults 125
Tourmaline **31**
Towers, stone **133**; *see also* Sea stacks
Trace fossils **119**
Traprock 74
Travertine **82**
Trilobites **105**
Tuff **71**
Turbidites 78

U

U-shaped valleys **171**
Ultraviolet light 48
Unconformity **149**

V

V-shaped valleys 171
Veins **127**
Ventifacts **157**
Volcanic bombs **139**
Volcanic necks or plugs **146**
Volcanoes
- cinder cones **144**
- stratovolcanoes 73
- *see also* Calderas; Lava; Lava tubes

W

Wackes. *see* Graywacke
Weathering and erosion
- arches **132**
- desert varnish **129**
- hoodoos **134**
- landslides **135**
- sheeting or exfoliation joints **131**
- spheroidal weathering **130**
- tafoni **128**
- towers and fins **133**

Windows, natural 132

X

Xenoliths 138

National Geographic
Pocket Guide to Rocks & Minerals
of North America

Prepared by the Book Division
Hector Sierra, *Senior Vice President and General Manager*
Janet Goldstein, *Senior Vice President and Editorial Director*
Jonathan Halling, *Creative Director*
Marianne Koszorus, *Design Director*
Susan Tyler Hitchcock, *Senior Editor*
R. Gary Colbert, *Production Director*
Jennifer A. Thornton, *Director of Managing Editorial*
Susan S. Blair, *Director of Photography*
Meredith C. Wilcox, *Director, Administration and Rights Clearance*

Staff for This Book
Barbara Payne, *Editor*
Paul Hess, *Text Editor*
Gail Spilsbury, *Project Editor*
Sanaa Akkach, *Art Director*
Katie Olsen, *Production Design Manager*
Noelle Weber, *Production Designer*
Catherine Herbert Howell, *Developmental Editor*
Miriam Stein, *Illustrations Editor*
Linda Makarov, *Designer*
Uliana Bazar, *Art Researcher*
Carl Mehler, *Director of Maps*
Marshall Kiker, *Associate Managing Editor*
Judith Klein, *Production Editor*
Galen Young, *Rights Clearance Specialist*

Production Services
Phillip L. Schlosser, *Senior Vice President*
Chris Brown, *Vice President, NG Book Manufacturing*
Nicole Elliott, *Director of Production*
George Bounelis, *Senior Production Manager*
Rachel Faulise, *Manager*
Robert L. Barr, *Manager*

National Geographic Partners
1145 17th Street N.W.
Washington, D.C. 20036-4688 U.S.A.

ISBN: 978-1-4262-1282-6

Printed in China

21/RRDH/5